Fritz

Mu

ller, William Sweetland Dallas

Facts and Arguments for Darwin

Fritz
Mu
̈
ller, William Sweetland Dallas
Facts and Arguments for Darwin
ISBN/EAN: 9783744638623
Printed in Europe, USA, Canada, Australia, Japan
Cover: Foto ©berggeist007 / pixelio.de

More available books at **www.hansebooks.com**

FACTS AND ARGUMENTS

FOR

DARWIN.

BY FRITZ MÜLLER.

WITH ADDITIONS BY THE AUTHOR.

TRANSLATED FROM THE GERMAN

By W. S. DALLAS, F.L.S.,

ASSISTANT SECRETARY TO THE GEOLOGICAL SOCIETY OF LONDON.

WITH ILLUSTRATIONS.

LONDON:

JOHN MURRAY, ALBEMARLE STREET.

1869.

A NATURALIST'S VOYAGE ROUND THE WORLD ;
being a JOURNAL OF RESEARCHES into the NATURAL HISTORY and GEOLOGY of
COUNTRIES VISITED. Post 8vo. 9s.

THE ORIGIN OF SPECIES, by MEANS of NATURAL
SELECTION ; or, THE PRESERVATION OF FAVOURED RACES IN THE STRUGGLE
FOR LIFE. Woodcuts. Post 8vo. 15s.

THE VARIOUS CONTRIVANCES by which BRITISH
and FOREIGN ORCHIDS are FERTILIZED by INSECTS, and on the GOOD
EFFECTS of INTERCROSSING. Woodcuts. Post 8vo. 9s.

THE VARIATION OF ANIMALS AND PLANTS
UNDER DOMESTICATION. Illustrations. 2 vols., 8vo. 28s.

SCIENCE

LONDON : PRINTED BY W. CLOWES AND SONS, DUKE STREET, STAMFORD STREET,
AND CHARING CROSS

TRANSLATOR'S PREFACE.

My principal reason for undertaking the translation of Dr. Fritz Müller's admirable work on the Crustacea, entitled 'Für Darwin,' was that it was still, although published as long ago as 1864, and highly esteemed by the author's scientific countrymen, absolutely unknown to a great number of English naturalists, including some who have occupied themselves more or less specially with the subjects of which it treats. It possesses a value quite independent of its reference to Darwinism, due to the number of highly interesting and important facts in the natural history and particularly the developmental history of the Crustacea, which its distinguished author, himself an unwearied and original investigator of these matters, has brought together in it. To a considerable section of English naturalists the tone adopted by the author in speaking of one of the greatest of their number will be a source of much gratification.

In granting his permission for the translation of his little book, Dr. Fritz Müller kindly offered to send some emendations and additions to certain parts of it. His notes included many corrections of printers' errors, some of which would have proved unintelligible without his aid, some small additions and notes which have been inserted in their proper places, and two longer pieces, one forming a foot-note near the close of Chap. XI. (p. 119), the other at the end of Chap. XII. (pp. 135–140), describing the probable mode of evolution of the Rhizocephala from the Cirripedia.

Of the execution of the translation I will say but little. My chief object in this, as in other cases, has been to furnish, as nearly as possible, a literal version of the original, regarding mere elegance of expression as of secondary importance in a scientific work. As much of Dr. Müller's German does not submit itself to such treatment very readily, I must beg his and the reader's indulgence for any imperfections arising from this cause.

W. S. D.

London, 15th Feb., 1869.

AUTHOR'S PREFACE.

IT is not the purpose of the following pages to discuss once more the arguments adduced for and against Darwin's theory of the origin of species, or to weigh them one against the other. Their object is simply to indicate a few facts favourable to this theory, collected upon the same South American ground, on which, as Darwin tells us, the idea first occurred to him of devoting his attention to "the origin of species,—that mystery of mysteries."

It is only by the accumulation of new and valuable material that the controversy will gradually be brought into a state fit for final decision, and this appears to be for the present of more importance than a repeated analysis of what is already before us. Moreover, it is but fair to leave it to Darwin himself at first to beat off the attacks of his opponents from the splendid structure which he has raised with such a master-hand.

F. M.

DESTERRO, 7th *Sept.*, 1863.

CONTENTS.

HISTORY OF CRUSTACEA.

CHAPTER I.

WHEN I had read Charles Darwin's book 'On the Origin of Species,' it seemed to me that there was one mode, and that perhaps the most certain, of testing the correctness of the views developed in it, namely, to attempt to apply them as specially as possible to some particular group of animals. Such an attempt to establish a genealogical tree, whether for the families of a class, the genera of a large family, or for the species of an extensive genus, and to produce pictures as complete and intelligible as possible of the common ancestors of the various smaller and larger circles, might furnish a result in three different ways.

1. In the first place, Darwin's suppositions when thus applied might lead to irreconcilable and contradictory conclusions, from which the erroneousness of the suppositions might be inferred. If Darwin's opinions are false, it was to be expected that contradictions would accompany their detailed application at every step, and

B

that these, by their cumulative force, would entirely destroy the suppositions from which they proceeded, even though the deductions derived from each particular case might possess little of the unconditional nature of mathematical proof.

2. Secondly, the attempt might be successful to a greater or less extent. If it was possible upon the foundation and with the aid of the Darwinian theory, to show in what sequence the various smaller and larger circles had separated from the common fundamental form and from each other, in what sequence they had acquired the peculiarities which now characterise them, and what transformations they had undergone in the lapse of ages,—if the establishment of such a genealogical tree, of a primitive history of the group under consideration, free from internal contradictions, was possible,—then this conception, the more completely it took up all the species within itself, and the more deeply it enabled us to descend into the details of their structure, must in the same proportion bear in itself the warrant of its truth, and the more convincingly prove that the foundation upon which it is built is no loose sand, and that it is more than merely "an intellectual dream."

3. In the third place, however, it was possible, and this could not but appear, *primâ facie*, the most probable case, that the attempt might be frustrated by the difficulties standing in its way, without settling the question, either way, in a perfectly satisfactory manner. But if it were only possible in this way to arrive for

oneself at a moderately certain independent judgment upon a matter affecting the highest questions so deeply, even this alone could not but be esteemed a great gain.

Having determined to make the attempt, I had in the first place to decide upon some particular class. The choice was necessarily limited to those the chief forms of which were easily to be obtained alive in some abundance. The Crabs and Macrurous Crustacea, the Stomapoda, the Diastylidæ, the Amphipoda and Isopoda, the Ostracoda and Daphnidæ, the Copepoda and Parasita, the Cirripedes and Rhizocephala of our coast, representing the class of Crustacea with the deficiency only of the Phyllopoda and Xiphosura, furnished a long and varied, and at the same time intimately connected series, such as was at my command in no other class. But even independently of this circumstance the selection of the Crustacea could hardly have been doubtful. Nowhere else, as has already been indicated by various writers, is the temptation stronger to give to the expressions " relationship, production from a common fundamental form," and the like, more than a mere figurative signification, than in the case of the lower Crustacea. Among the parasitic Crustacea, especially, everybody has long been accustomed to speak, in a manner scarcely admitting of a figurative meaning, of their arrest of development by parasitism, as if the transformation of species were a matter of course. It would certainly never appear to any one to be a pastime worthy of the Deity, to amuse himself with the contrivance of these marvellous cripplings, and so they were supposed to have fallen by their own fault, like Adam, from their previous state of perfection.

That a great part of the larger and smaller groups into which this class is divided, might be regarded as satisfactorily established, was a further advantage not to be undervalued; whilst in two other classes with which I was familiar, namely, the Annelida and Acalephæ, all the attempted arrangements could only be considered preliminary revisions. These undisplaceable groups, like the sharply marked forms of the hard, many-jointed dermal framework, were not only important as safe starting points and supports, but were also of the highest value as inflexible barriers in a problem in which, from its very nature, fancy must freely unfold her wings.

When I thus began to study our Crustacea more closely from this new stand-point of the Darwinian theory,—when I attempted to bring their arrangements into the form of a genealogical tree, and to form some idea of the probable structure of their ancestors,—I speedily saw (as indeed I expected) that it would require years of preliminary work before the essential problem could be seriously handled. The extant systematic works generally laid more weight upon the characters separating the genera, families and orders, than upon those which unite the members of each group, and consequently often furnished but little employable material. But above all things a thorough knowledge of development was indispensable, and every one knows how imperfect is our present knowledge of this subject. The existing deficiencies were the more difficult to supply, because, as Van Beneden remarks with regard to the Decapoda, from the often incredible difference in the development of the most nearly allied forms, these must be separately studied—usually family by family, and frequently genus by genus—nay, sometimes, as in the case of *Penëus*, even species by species; and because these investigations, in themselves troublesome and tedious, often depend for their success upon a lucky chance.

But although the satisfactory completion of the "Genealogical tree of the Crustacea" appeared to be an undertaking for which the strength and life of an individual would hardly suffice, even under more favourable circumstances than could be presented by a distant island, far removed from the great market of scientific life, far from libraries and museums—nevertheless its practicability became daily less doubtful in my eyes, and fresh observations daily made me more favourably inclined towards the Darwinian theory.

In determining to state the arguments which I derived from the consideration of our Crustacea in favour of Darwin's views, and which (together with more general considerations and observations in other departments), essentially aided in making the correctness of those views seem more and more palpable to me, I am chiefly influenced by an expression of Darwin's : "Whoever," says he ('Origin of Species,' p. 482), "is led to believe that species are mutable, will do a good service by conscientiously expressing his conviction." To the

desire expressed in these words I respond, for my own
part, with the more pleasure, as this furnishes me with
an opportunity of publicly giving expression in words to
the thanks which I feel most deeply to be due from me
to Darwin for the instructions and suggestions for which
I am so deeply indebted to his book. Accordingly I
throw this sand-grain with confidence into the scale
against "the load of prejudice by which this subject is
overwhelmed," without troubling myself as to whether
the priests of orthodox science will reckon me amongst
dreamers and children in knowledge of the laws of
nature.

CHAPTER II.

THE SPECIES OF MELITA.

A FALSE supposition, when the consequences proceeding
from it are followed further and further, will sooner or
later lead to absurdities and palpable contradictions.
During the period of tormenting doubt—and this was
by no means a short one—when the pointer of the
scales oscillated before me in perfect uncertainty between the *pro* and the *con*, and when any fact leading
to a quick decision would have been most welcome
to me, I took no small pains to detect some such contradictions among the inferences as to the class of
Crustacea furnished by the Darwinian theory. But I
found none, either then, or subsequently. Those which
I thought I had found were dispelled on closer consideration, or actually became converted into supports
for Darwin's theory.

Nor, so far as I am aware, have any of the *necessary*
consequences of Darwin's hypotheses been proved by any
one else, to stand in clear and irreconcilable contradiction. And yet, as the most profound students of the
animal kingdom are amongst Darwin's opponents, it
would seem that it ought to have been an easy matter
for them to crush him long since beneath a mass of ab-

surd and contradictory inferences, if any such were to be drawn from his theory. To this want of demonstrated contradictions I think we may ascribe just the same importance in Darwin's favour, that his opponents have attributed to the absence of demonstrated intermediate forms between the species of the various strata of the earth. Independently of the reasons which Darwin gives for the preservation of such intermediate forms being only exceptional, this last mentioned circumstance will not be regarded as of very great significance by any one who has traced the development of an animal upon larvæ fished from the sea, and had to seek in vain for months, and even years, for those transitional forms, which he nevertheless knew to be swarming around him in thousands.

A few examples may show how contradictions might come forth as necessary results of the Darwinian hypotheses.

It seems to be a necessity for all crabs which remain for a long time out of the water (but why is of no consequence to us here), that air shall penetrate from behind into the branchial cavity. Now these crabs, which have become more or less estranged from the water, belong to the most different families—the Raninidæ (*Ranina*), Eriphinæ (*Eriphia gonagra*), Grapsoidæ (*Aratus, Sesarma,* &c.), Ocypodidæ (*Gelasimus, Ocypoda*), &c., and the separation of these families must doubtless be referred to a much earlier period than the habit of leaving the water displayed by some of their members. The arrangements connected with aerial respiration, therefore, could not be inherited from a common ancestor, and could scarcely be accordant in their construction. If there were any such accordance not referable to accidental resemblance among them, it would have to be laid in the scale as evidence against the correctness of Darwin's views. I shall show hereafter how in this case the result, far from presenting such contradictions, was rather in the most complete harmony with what might be predicted from Darwin's theory.

A second example.—We are already acquainted with four species of *Melita* (*M. valida, setipes, anisochir,* and

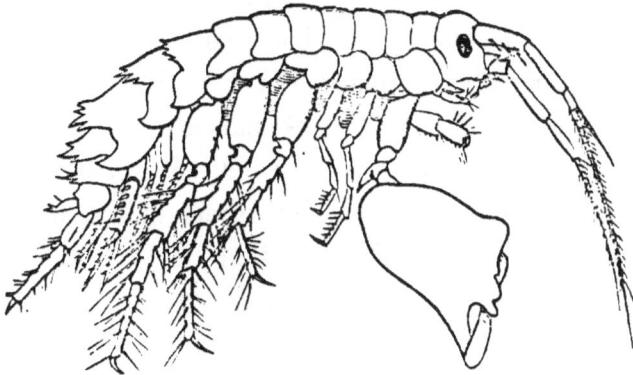

Fig. 1.[1]

Fresnelii), and I can add a fifth (fig. 1), in which the second pair of feet bears upon one side a small hand of the usual structure, and on the other an enormous clasp-forceps. This want of symmetry is something so

[1] Fig. 1. *Melita exilii* n. sp., male, enlarged five times. The large branchial lamellæ are seen projecting between the legs.

unusual among the Amphipoda, and the structure of
the clasp-forceps differs so much from what is seen elsewhere in this order, and agrees so closely in the five
species, that one must unhesitatingly regard them as
having sprung from common ancestors belonging to
them alone among known species. But one of these
species, *M. Fresnelii*, discovered by Savigny, in Egypt,
is said to want the secondary flagellum of the anterior
antennæ, which occurs in the others. From the trust-
worthiness of all Savigny's works there can scarcely be
a doubt as to the correctness of this statement. Now,
if the presence or absence of the secondary flagellum
possessed the significance of a distinctive generic character, which is usually ascribed to it, or if there were
other important differences between *Melita Fresnelii*
and the other species above-mentioned, which would
make it seem natural to separate *M. Fresnelii* as a distinct genus, and to leave the others united with the
rest of the species of *Melita*—that is to say, in the
sense of the Darwinian theory, if we assume that all
the other *Melitæ* possessed common ancestors, which
were not at the same time the ancestors of *M. Fresnelii*—this would stand in contradiction to the conclusion, derived from the structure of the clasp-forceps,
that *M. Fresnelii* and the four other species above-mentioned possessed common ancestors, which were not also
the ancestors of the remaining species of *Melita*. It
would follow

From the structure of the clasp-forceps:

From the presence or absence of the secondary flagellum.

M. palmata, &c. *M. exilii*, &c. *M. Fresnelii*. | *M. palmata*, &c. *M. exilii*, &c. *M. Fresnelii*.

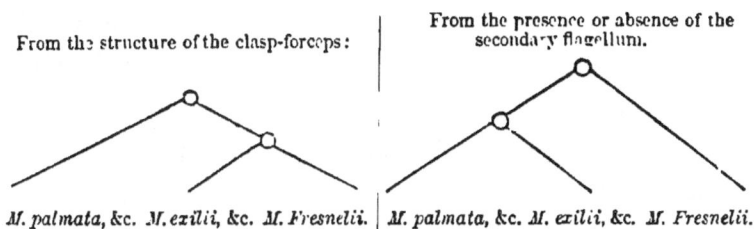

As, in the first case, among the Crabs, a typical agreement of arrangements produced independently of each other would have been a very suspicious circumstance for Darwin's theory, so also, in the second, would any difference more profound than that of very nearly allied species. Now it seems to me that the secondary flagellum can by no means furnish a reason for doubting the close relationship of *M. Fresnelii* to *M. exilii*, &c., which is indicated by the peculiar structure of the unpaired clasp-forceps. In the first place we must consider the possibility that the secondary flagellum, which is not always easy to detect, may only have been overlooked by Savigny, as indeed Spence Bate supposes to have been the case. If it is really deficient it must be remarked that I have found it in species of the genera *Leucothoë*, *Cyrtophium* and *Amphilochus*, in which genera it was missed by Savigny, Dana and Spence Bate—that a species proved by the form of the epimera (*coxæ* Sp. B.) of the caudal feet (*uropoda* Westw.), &c., to be a true *Amphithoë*[2] possesses it—that in many species of *Cerapus* it is reduced to a scarcely perceptible

[2] I accept this and all the other genera of Amphipoda here mentioned, with the limits given to them by Spence Bate ('Catal. of Amphipodous Crustacea').

rudiment—nay, that it is sometimes present in youth and disappears (although perhaps not without leaving some trace) at maturity, as was found by Spence Bate to be the case in *Acanthonotus Owenii* and *Atylus carinatus*, and I can affirm with regard to an *Atylus* of these seas, remarkable for its plumose branchiæ—and that from all this, at the present day when the increasing number of known Amphipoda and the splitting of them into numerous genera thereby induced, compels us to descend to very minute distinctive characters, we must nevertheless hesitate before employing the secondary flagellum as a generic character. The case of *Melita Fresnelii* therefore cannot excite any doubts as to Darwin's theory.

CHAPTER III.

MORPHOLOGY OF CRUSTACEA—NAUPLIUS-LARVÆ.

IF the absence of contradictions among the inferences deduced from them for a narrow and consequently easily surveyed department must prepossess us in favour of Darwin's views, it must be welcomed as a positive triumph of his theory if far-reaching conclusions founded upon it should *subsequently* be confirmed by facts, the existence of which science, in its previous state, by no means allowed us to suspect. From many results of this kind upon which I could report, I select as examples, two, which were of particular importance to me, and relate to discoveries the great significance of which in the morphology and classification of the Crustacea will not be denied even by the opponents of Darwin.

Considerations upon the developmental history of the Crustacea had led me to the conclusion that, if the higher and lower Crustacea were at all derivable from common progenitors, the former also must once have passed through Nauplius-like conditions. Soon afterwards I discovered Naupliiform larvæ of Shrimps ('Archiv für Naturg.' 1860, i. p. 8), and I must admit that this discovery gave me the first decided turn in Darwin's favour.

The similar number of segments [1] occurring in the

[1] Like Claus I do not regard the eyes of the Crustacea as limbs, and therefore admit no ocular segment ; on the other hand I count in the median piece of the tail, to which the character of a segment is often denied. In opposition to its interpretation as a segment of the body, only the want of limbs can be cited ; in its favour we have the relation of the intestine, which usually opens in this piece, and sometimes even traverses its whole length, as in *Microdeutopus* and some other Amphipoda. In *Microdeutopus*, as Spence Bate has already pointed out, one is even led to regard small processes of this tubular caudal piece as rudimentary members. Bell also (' Brit. Stalk-eyed Crust.' p. xx.), states that he observed limbs of the last segment in *Palæmon serratus* in the form of small movable points.

The attempt has often been made to divide the body of the higher Crustacea into small sections composed of equal numbers of segments, these sections consisting of 3, 5 or 7 segments. None of these attempts has ever met with general acceptance ; my own investigations lead me to a conception which nearly approaches Van Beneden's. I assume four sections of 5 segments each—the primitive body, the fore-body, the hind-body, and the middle-body. The primitive body includes the segments which the naupliiform larva brings with it out of the egg ; it is afterwards divided, by the younger sections which become developed in its middle, into the head and tail. To this primitive body belong the two pairs of antennæ, the mandibles and the caudal feet (" posterior pair of pleopoda," Sp. B.). Even in the mature animal the fact that these terminal sections belong to one another is sometimes betrayed by the resemblance of their appendages, especially that of the outer branch of the caudal feet, with the outer branch (the so-called scale) of the second pair of antennæ. Like the antennæ, the caudal feet may also become the bearers of high sensorial apparatus, as is shown by the ear of *Mysis*.

The sequence of the sections of the body in order of time seems originally to have been, that first the fore-body, then the hind-body, and finally the middle-body was formed. The fore-body appears, in the adult animal, to be entirely or partially amalgamated with the head ; its appendages (*siagonopoda* Westw.) are all or in part serviceable for the reception of food, and generally sharply distinguished from those of the following group. The segments of the middle-body seem always

Crabs and Macrura, Amphipoda and Isopoda, in which the last seven segments are always different from the preceding ones in the appendages with which they are furnished, could only be regarded as an inheritance from the same ancestors. And if at the present day the majority of the Crabs and Macrura, and indeed the Stalk-eyed Crustacea in general, pass through Zoëa-like developmental states, and the same mode of transformation was to be ascribed to their ancestors, the same thing must also apply, if not to the immediate ancestors of the Amphipoda and Isopoda, at least to the common progenitors of these and the Stalk-eyed Crustacea. Any such assumption as this was, however, very hazardous, so long as not a single fact properly relating to the

to put forth limbs immediately after their own appearance, whilst the segments of the hind-body often remain destitute of feet through long portions of the larval life or even throughout life (as in many female Diastylidæ), a reason, among many others, for not, as is usual, regarding the middle-body of the Crustacea as equivalent to the constantly footless abdomen of Insects. The appendages of the middle-body (*pereiopoda*) seem never, even in their youngest form, to possess two equal branches, a peculiarity which usually characterises the appendages of the hind-body. This is a circumstance which renders very doubtful the equivalence of the middle-body of the Malacostraca with the section of the body which in the Copepoda bears the swimming feet and in the Cirripedia the cirri.

The comprehension of the feet of the hind-body and tail in a single group (as "fausses pattes abdominales," or as " pleopoda ") seems not to be justifiable. When there is a metamorphosis, they are probably always produced at different periods, and they are almost always quite different in structure and function. Even in the Amphipoda, in which the caudal feet usually resemble in appearance the last two pairs of abdominal feet, they are in general distinguished by some sort of peculiarity, and whilst the abdominal feet are reproduced in wearisome uniformity throughout the entire order, the caudal feet are, as is well known, amongst the most variable parts of the Amphipoda.

Edriophthalma could be adduced in its support, as the
structure of this very coherent group seemed to be
almost irreconcilable with many peculiarities of the
Zoëa. Thus, in my eyes, this point long constituted one
of the chief difficulties in the application of the Darwinian views to the Crustacea, and I could scarcely
venture to hope that I might yet find traces of this
passage through the Zoëa-form among the Amphipoda
or Isopoda, and thus obtain a positive proof of the correctness of this conclusion. At this point Van Beneden's statement that a cheliferous Isopod (*Tanais
Dulongii*), belonging, according to Milne-Edwards,
to the same family as the common *Asellus aqua-
ticus*, possesses a carapace like the Decapoda, directed
my attention to these animals, and a careful exa-

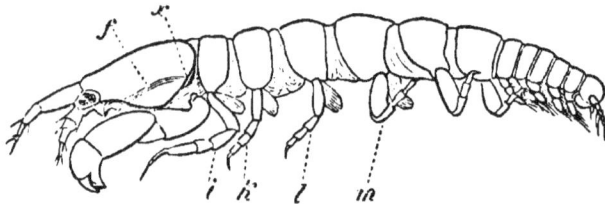

Fig. 2.[2]

mination proved that these Isopods have preserved,
more truly than any other adult Crustacea, many of the
most essential peculiarities of the Zoëæ, especially their

---

[2] *Tanais dubius* (?) Kr. ♀, magnified 25 times, showing the orifice
of entrance (*x*) into the cavity overarched by the carapace, in which
an appendage of the second pair of maxillæ (*f*) plays. On four feet
(*i, k, l, m*) are the rudiments of the lamellæ which subsequently form
the brood-cavity.

mode of respiration. Whilst in all other Oniscoida the abdominal feet serve for respiration, these in our cheliferous Isopod (fig. 2) are solely motory organs, into which no blood-corpuscle ever enters, and the chief seat of respiration is, as in the *Zoëæ*, in the lateral parts of the carapace, which are abundantly traversed by currents of blood, and beneath which a constant stream of water passes, maintained, as in *Zoëæ* and the adult Decapoda, by an appendage of the second pair of maxillæ, which is wanting in all other Edriophthalma.

For both these discoveries, it may be remarked in passing, science is indebted less to a happy chance than immediately to Darwin's theory.

Species of *Penëus* live in the European seas, as well as here, and their *Nauplius*-brood has no doubt repeatedly passed unnoticed through the hands of the numerous naturalists who have investigated those seas, as well as through my own,[3] for it has nothing which could attract particular attention amongst the multifarious and often wonderful *Nauplius*-forms. When I, fancying from the similarity of its movements that it was a young *Penëus-Zoëa*, had for the first time captured such a larva, and on bringing it under the microscope found a *Nauplius* differing *toto cœlo* from this *Zoëa*, I might have thrown it aside as being completely foreign to the developmental series which I was tracing, if the idea of early Naupliiform stages of the higher Crustacea, which in-

[3] Mecznikow has recently found Naupliiform shrimp-larvæ in the sea near Naples.

deed I did not believe to be still extant, had not at the moment vividly occupied my attention.

And if I had not long been seeking among the Edriophthalma for traces of the supposititious _Zoëa_-state, and seized with avidity upon everything that promised to make this refractory Order serviceable to me, Van Beneden's short statement could hardly have affected me so much in the manner of an electric shock, and impelled me to a renewed study of the _Tanaides_, especially as I had once before plagued myself with them in the Baltic, without getting any further than my predecessors, and I have not much taste for going twice over the same ground.

CHAPTER IV.

OUR *Tanais*, which in nearly all the particulars of its structure is an extremely remarkable animal, furnished me with a second fact worthy of notice in connection with the theory of the origin of species by natural selection.

When handlike or cheliform structures occur in the Crustacea, these are usually more strongly developed in the males than in the females, often becoming enlarged in the former to quite a disproportionate size, as we have already seen to be the case in *Melita*. A better known example of such gigantic chelæ is presented by the males of the Calling Crabs (*Gelasimus*), which are said in running to carry these claws "elevated, as if beckoning with them"—a statement which, however, is not true of all the species, as a small and particularly large-clawed one, which I have seen running about by thousands in the cassava-fields at the mouth of the Cambriú, always holds them closely pressed against its body.

A second peculiarity of the male Crustacea consists not unfrequently in a more abundant development on

c 2

the flagellum of the anterior antennæ of delicate filaments which Spence Bate calls "auditory cilia," and
which I have considered to be olfactory organs, as
did Leydig before me, although I was not aware of
it. Thus they form long dense tufts in the males of
many Diastylidæ, as Van Beneden also states with regard
to *Bodotria*, whilst the females only possess them more
sparingly. In the Copepoda, Claus called attention to
the difference of the sexes in this respect. It seems to
me, as I may remark in passing, that this stronger development in the males is greatly in favour of the opinion
maintained by Leydig and myself, as in other cases male
animals are not unfrequently guided by the scent in
their pursuit of the ardent females.

Now, in our *Tanais*, the young males up to the last
change of skin preceding sexual maturity resemble the
females, but then they undergo an important metamorphosis. Amongst other things they lose the moveable
appendages of the mouth even to those which serve for
the maintenance of the respiratory current; their intestine is always found empty, and they appear only to live
for love. But what is most remarkable is, that they now
appear under two different forms. Some (fig. 3) acquire
powerful, long-fingered, and very mobile chelæ, and,
instead of the single olfactory filament of the female,
have from 12 to 17 of these organs, which stand two or
three together on each joint of the flagellum. The others
(fig. 5) retain the short thick form of the chelæ of the
females ; but, on the other hand, their antennæ (fig. 6)
are equipped with a far greater number of olfactory

filaments, which stand in groups of from five to seven together.

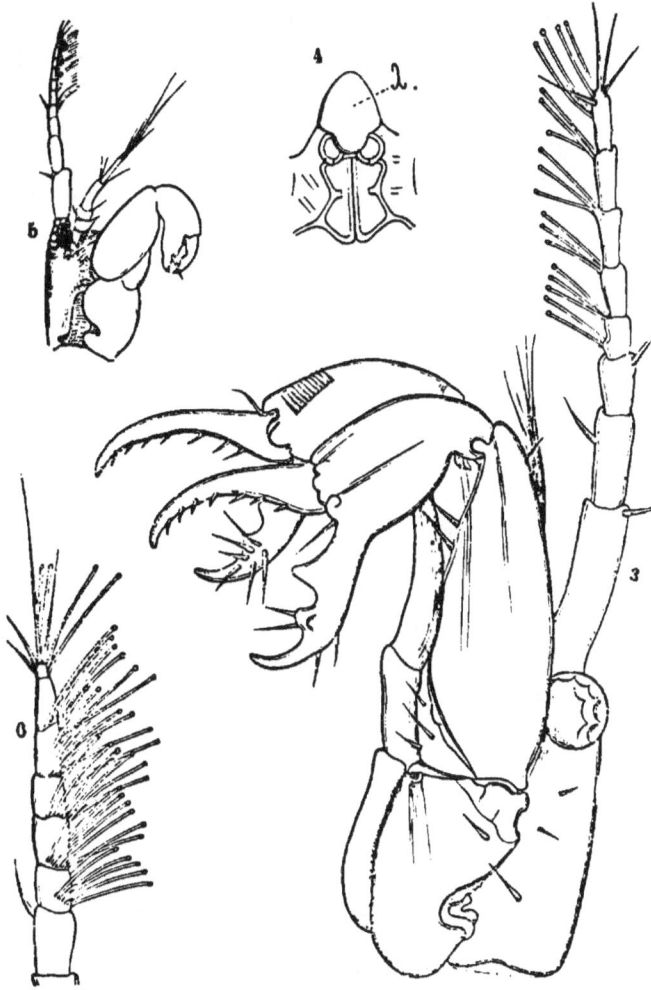

Figs. 3-6.[1]

[1] Fig. 3. Head of the ordinary form of the male of *Tanais dubius* (?) Kr. magn. 90 times. The terminal setæ of the second pair of antennæ project between the cheliferous feet. Fig. 4. Buccal region of the same from below; λ, labrum. Fig. 5. Head of the rarer form of the male, magn. 25 times. Fig. 6. Flagellum of the same, with olfactory filaments, magn. 90 times.

In the first place, and before inquiring into its significance, I will say a word upon this fact itself. It was
natural to consider whether two different species with
very similar females and very different males might not
perhaps live together, or whether the males, instead of
occurring in two sharply defined forms, might not be
only variable within very wide limits. I can admit
neither of these suppositions. Our *Tanais* lives among
densely interwoven Confervæ, which form a coat of
about an inch in thickness upon stones in the neighbourhood of the shore. If a handful of this green felt is put
into a large glass with clear sea-water, the walls of the
glass are soon seen covered with hundreds, nay with
thousands, of these little, plump, whitish Isopods. In
this way I have examined thousands of them with the
simple lens, and I have also examined many hundreds
with the microscope, without finding any differences
among the females, or any intermediate forms between
the two kinds of males.

To the old school this occurrence of two kinds of
males will appear to be merely a matter of curiosity.
To those who regard the "plan of creation" as the
"free conception of an Almighty intellect, matured
in the thoughts of the latter before it is manifested in
palpable, external forms," it will appear to be a mere
caprice of the Creator, as it is inexplicable either
from the point of view of practical adaptation, or
from the "typical plan of structure." From the side
of Darwin's theory, on the contrary, this fact acquires
meaning and significance, and it appears in return

to be fitted to throw light upon a question in which
Bronn saw "the first and most material objection
against the new theory," namely, how it is possible
that from the accumulation in various directions of the
smallest variations running out of one another, varieties
and species are produced, which stand out from the primary form clearly and sharply like the petiolated leaf
of a Dicotyledon, and are not amalgamated with the
primary form and with each other like the irregular
curled lobes of a foliaceous Lichen.

Let us suppose that the males of our *Tanais*, hitherto
identical in structure, begin to vary, in all directions as
Bronn thinks, for aught I care. If the species was
adapted to its conditions of existence, if the *best* in this
respect had been attained and secured by natural selection, fresh variations affecting the species as a species
would be retrogressions, and thus could have no prospect
of prevailing. They must rather have disappeared
again as they arose, and the lists would remain open to
the males under variation, only in respect of their sexual
relations. In these they might acquire advantages over
their rivals by their being enabled either to seek or to
seize the females better. The best smellers would over-
come all that were inferior to them in this respect, unless
the latter had other advantages, such as more powerful
chelæ, to oppose to them. The best claspers would over-
come all less strongly armed champions, unless these
opposed to them some other advantage, such as sharper
senses. It will be easily understood how in this manner all
the intermediate steps less favoured in the development

of the olfactory filaments or of the chelæ would disappear
from the lists, and two sharply defined forms, the best
smellers and the best claspers, would remain as the sole
adversaries. At the present day the contest seems to
have been decided in favour of the latter, as they occur
in greatly preponderating numbers, perhaps a hundred
of them to one smeller.

To return to Bronn's objection. When he says that
"for the support of the Darwinian theory, and in order
to explain why many species do not coalesce by means
of intermediate forms, he would gladly discover some
external or internal principle which should compel the
variations of each species to advance in *one* direction,
instead of merely permitting them in all directions," we
may, in this as in many other cases, find such a principle
in the fact that actually only a few directions stand open
in which the variations are at the same time improve-
ments, and in which therefore they can accumulate and
become fixed; whilst in all others, being either indifferent
or injurious, they will go as lightly as they come.

The occurrence of two kinds of males in the same
species may perhaps not be a very rare phenomenon
in animals in which the males differ widely from
the females in structure. But only in those which
can be procured in sufficient abundance, will it be
possible to arrive at a conviction that we have not
before us either two different species, or animals of
different ages. From my own observation, although
not very extensive, I can give a second example. It
relates to a shore-hopper (*Orchestia*). The animal

(fig. 7) lives in marshy places in the vicinity of the sea, under decaying leaves, in the loose earth which the Marsh Crabs (*Gelasimus, Sesarma, Cyclograpsus*, &c.)

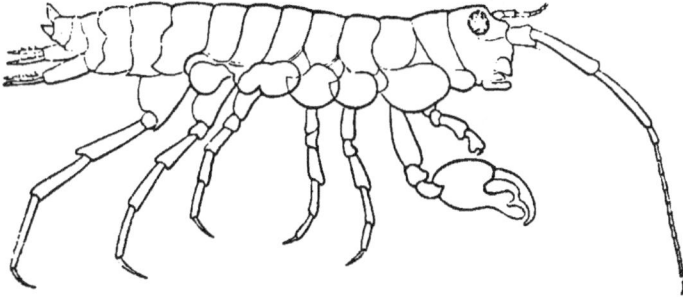

Fig. 7.[2]

throw up around the entrance to their burrows, and even under dry cow-dung and horse-dung. If this species removes to a greater distance from the shore than the majority of its congeners (although some of them advance very far into the land and even upon mountains of a thousand feet in height, such as *O. tahitensis, telluris*, and *sylvicola*), its male differs still more from all known species by the powerful chelæ of the second pair of feet. *Orchestia gryphus*, from the sandy coast of Mönchgut, alone presents a somewhat similar structure; but in a far less degree ; elsewhere the form of the hand usual in the Amphipoda occurs. Now there is a considerable difference between the males of this species, especially in the structure of these chelæ—a difference so great that we can scarcely find a parallel to it elsewhere between two species of the genus— and yet, as in *Tanais*, we do not meet with a

[2] Fig. 7. *Orchestia Darwinii*, n. sp. male.

long series of structures running into one another, but only two forms united by no intermediate terms (figs. 8 and 9). The males would be unhesitatingly regarded

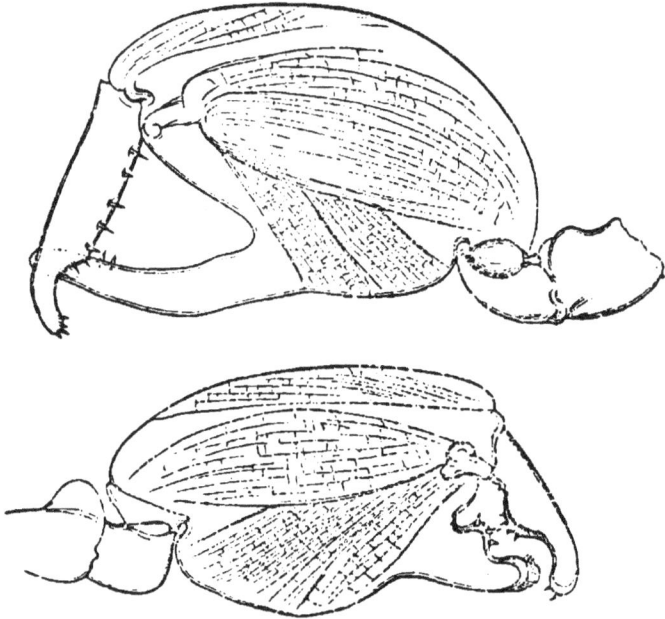

Figs. 8 and 9. [3]

as belonging to two well-marked species if they did not live on the same spot, with undistinguishable females. That the two forms of the chelæ of the males occur in this species is so far worthy of notice, because the formation of the chelæ, which differs widely from the ordinary structure in the other species, indicates that it has quite recently undergone considerable changes, and therefore such a phenomenon was to be expected in it rather than in other species.

[3] Figs. 8 and 9. The two forms of the chelæ of the male of *Orchestia Darwinii*, magn. 45 times.

I cannot refrain from taking this opportunity of remarking that (so far as appears from Spence Bate's
catalogue), for two different kinds of males (*Orchestia
telluris* and *sylvicola*) which live together in the forests
of New Zealand, only one form of female is known, and
hazarding the supposition that we have here a similar
case. It does not seem to me to be probable that two
nearly allied species of these social Amphipoda should
occur mixed together under the same conditions of life.

As the males of several species of *Melita* are distinguished by the powerful unpaired clasp-forceps, the
females of some
other species of the
same genus are
equally distinguish-
ed from all other
Amphipoda by the
circumstance that
in them a peculiar
apparatus is de-
veloped which fa-
cilitates their being
held by the male.
The coxal lamellæ

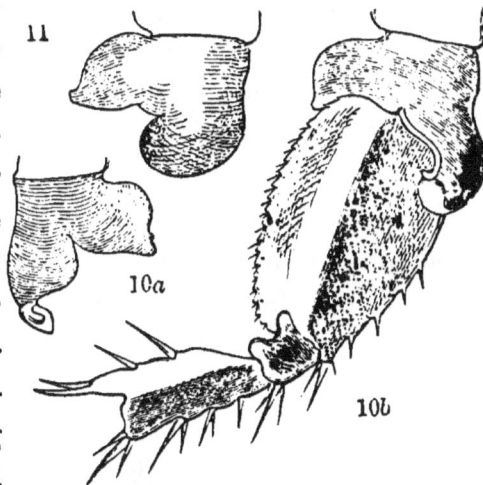

Figs. 10 and 11.[4]

of the penultimate pair of feet are produced into hook-
like processes, of which the male lays hold with the

---

[4] Fig. 10. Coxal lamella of the penultimate pair of feet of the male (*a*),
and coxal lamella, with the three following joints of the same pair of
feet of the female (*b*) of *Melita Messalina*, magn. 45 diam.

Fig. 11. Coxal lamella of the same pair of feet of the female of
M. insatiabilis.

hands of the first pair of feet. The two species in
which I am acquainted with this structure are amongst
the most salacious animals of their order, even females
which are laden with eggs in all stages of development,
not unfrequently have their males upon their backs.
The two species are nearly allied to *Melita palmata*
Leach (*Gammarus Dugesii*, Edw.), which is widely
distributed on the European coasts, and has been frequently investigated; unfortunately, however, I can
find no information as to whether the females of this or
any other European species possess a similar contrivance. In *M. exilii* all the coxal lamellæ are of the
ordinary formation. Nevertheless, be this as it will,
whether they exist in two or in twenty species, the
occurrence of these peculiar hook-like processes is
certainly very limited.

Now our two species live sheltered beneath slightly
tilted stones in the neighbourhood of the shore: one of
them, *Melita Messalina*, so high that it is but rarely
covered by the water; the other, *Melita insatiabilis*, a
little lower; both species live together in numerous
swarms. We cannot therefore suppose that the loving
couples are threatened with disturbance more frequently
than those of other species, nor would it be more difficult
for the male, than for those of other species, in case of
his losing his female, to find a new one. Nor is it any
more easy to see how the contrivance on the body of
the female for insuring the act of copulation could be
injurious to other species. But so long as it is not
demonstrated that our species are particularly in want

of this contrivance, or that the latter would rather be injurious than beneficial to other species, its presence only in these few Amphipoda will have to be regarded not as the work of far-seeing wisdom, but as that of a favourable chance made use of by Natural Selection. Under the latter supposition its isolated occurrence is intelligible, whilst we cannot perceive why the Creator blessed just these few species with an apparatus which he found to be quite compatible with the "general plan of structure" of the Amphipoda, and yet denied it to others which live under the same external conditions, and equal them even in their extraordinary salacity. Associated with, or in the immediate vicinity of the two species of *Melita*, live two species of *Allorchestes*, the pairs of which are met with almost more numerously than the single animals, and yet their females show no trace of the above-mentioned processes of the coxal lamellæ.

These cases, I think, must be brought to bear against the conception supported with so much genius and knowledge by Agassiz, that species are embodied thoughts of the Creator; and, with these, all similar instances in which arrangements which would be equally beneficial to all the species of a group are wanting in the majority and only conferred upon a few special favourites, which do not seem to want them any more than the rest.

CHAPTER V.

RESPIRATION IN LAND CRABS.

AMONG the numerous facts in the natural history of
the Crustacea upon which a new and clear light is
thrown by Darwin's theory, besides the two forms of
the males in our *Tanais* and in *Orchestia Darwinii*,
there is one which appears to me of particular importance, namely, the character of the branchial cavity
in the air-breathing Crabs, of which, unfortunately, I
have been unable to investigate some of the most
remarkable (*Gecarcinus, Ranina*). As this character,
namely, the existence of an entrance behind the
branchiæ, has hitherto been noticed, even as a fact,
only in *Ranina*, I will go into it in some detail. I
have already mentioned that, as indeed is required by
Darwin's theory, this entrant orifice is produced in
different manners in the different families.

In the Frog-crab (*Ranina*) of the Indian Ocean,
which, according to Rumphius, loves to climb up on
the roofs of the houses, the ordinary anterior entrant
orifice is entirely wanting according to Milne-Edwards,
and the entrance of a canal opening into the hindmost
parts of the branchial cavity is situated beneath the
commencement of the abdomen.

The case is most simple in some of the Grapsoidæ, as in *Aratus Pisonii*, a charming, lively Crab which ascends the mangrove bushes (*Rhizophora*) and gnaws their leaves. By means of its short but remarkably acute claws, which prick like pins when it runs over the hand, this Crab climbs with the greatest agility upon the thinnest twigs. Once, when I had one of these animals sitting upon my hand, I noticed that it elevated the hinder part of its carapace, and that by this means a wide fissure was opened upon each side above the last pair of feet, through which I could look far into the branchial cavity. I have since been unable to procure this remarkable animal again, but on the other hand, I have frequently repeated the same observation upon another animal of the same family (apparently a true *Grapsus*), which lives abundantly upon the rocks of our coast. Whilst the hinder part of the carapace rises and the above-mentioned fissure is formed, the anterior part seems to sink, and to narrow or entirely close the anterior entrant orifice. Under water the elevation of the carapace never takes place. The animal therefore opens its branchial cavity in front or behind, according as it has to breathe water or air. How the elevation of the carapace is effected I do not know, but I believe that a membranous sac, which extends from the body cavity far into the branchial cavity beneath the hinder part of the carapace, is inflated by the impulsion of the fluids of the body, and the carapace is thereby raised.

I have also observed the same elevation of the carapace in some species of the allied genera *Sesarma* and
Cyclograpsus, which dig deep holes in marshy ground,
and often run about upon the wet mud, or sit, as if
keeping watch, before their burrows. One must, however, wait for a long time with these animals, when
taken out of the water, before they open their branchial
cavity to the air, for they possess a wonderful arrangement, by means of which they can continue to breathe
water for some time when out of the water. The
orifices for the egress of the water which has served for
respiration, are situated in these, as in most Crabs, in
the anterior angles of the buccal frame (" cadre buccal,"
M.-Edw.), whilst the entrant fissures of the branchial
cavity extend from its hinder angles above the first
pair of feet. Now that portion of the carapace which
extends at the sides of the mouth between the two
orifices (" régions ptérygostomiennes "), appears in our
animals to be divided into small square compartments.
Milne-Edwards has already pointed this out as a particularly remarkable peculiarity. This appearance is
caused partly by small wart-like elevations, and partly
and especially by curious geniculated hairs, which to a
certain extent constitute a fine net or hair-sieve extended
immediately over the surface of the carapace. Thus
when a wave of water escapes from the branchial
cavity, it immediately becomes diffused in this network
of hairs and then again conveyed back to the branchial
cavity by vigorous movements of the appendage of the

outer maxilliped which works in the entrant fissure. Whilst the water glides in this way over the carapace in the form of a thin film, it will again saturate itself with oxygen, and may then serve afresh for the purposes of respiration. In order to complete this arrangement the outer maxillipeds, as indeed has long been known, bear a projecting ridge furnished with a dense fringe of hairs, which commences in front near their median line and passes backwards and outwards to the hinder angle of the buccal frame. Thus the two ridges of the right and left sides form together a triangle with the apex turned forwards,—a breakwater by which the water flowing from the branchial cavity is kept away from the mouth and reconducted to the branchial cavity. In very moist air the store of water contained in the branchial cavity may hold out for hours, and it is only when this is used up that the animal elevates its carapace in order to allow the air to have access to its branchiæ from behind.

In *Eriphia gonagra* the entrant orifices of the respiratory cavity serving for aerial respiration are situated, not, as in the Grapsoidæ, above, but behind the last pair of feet at the sides of the abdomen.

The swift-footed Sand-Crabs (*Ocypoda*) are exclusively terrestrial animals, and can scarcely live for a single day in water; in a much shorter period a state of complete relaxation occurs and all voluntary movements cease.[1] In these a peculiar arrangement

---

[1] As this was not observed in the sea, but in glass vessels containing sea-water, it might be supposed that the animals become exhausted

on the feet of the third and fourth pairs (fig. 12) has long been known, although its connexion with the

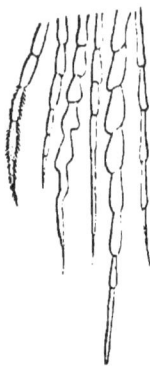

branchial cavity has not been suspected. These two pairs of feet are more closely approximated than the rest; the opposed surfaces of their basal joints (therefore the hinder surface on the third, and the anterior surface on the fourth feet) are smooth and polished, and their margins bear a dense border of long, silky, and peculiarly formed hairs (fig. 13). Milne-Edwards who rightly compares these surfaces, as to their appearance, with articular surfaces, thinks that they serve to diminish the friction between the two feet. In considering this interpretation, the question could not

Fig. 13.[3]   Fig. 12.[2]

and die, not because they are under water but because they have consumed all the oxygen which it contained. I therefore put into the same water from which I had just taken an unconscious *Ocypoda*, with its legs hanging loosely down, a specimen of *Lupea diacantha* which had been reduced to the same state by being kept in the air, and this recovered in the water just as the *Ocypoda* did in the air.

[2] Fig. 12. Posterior entrance to the branchial cavity of *Ocypoda rhombea*, Fab., nat. size. The carapace and the fourth foot of the right side are removed.

[3] Fig. 13. Points of some of the hairs of the basal joints of the foot, magn. 45 diam.

but arise why such an arrangement for the diminution of friction should be necessary in these particular Crabs and between these two feet, leaving out of consideration the fact that the remarkable brushes of hair, which on the other hand must increase friction, also remain unexplained. But as I was bending the feet of a large Sand-Crab to and fro in various directions, in order to see in what movements of the animal friction occurred at the place indicated, and whether these might, perhaps, be movements of particular importance to it and such as would frequently recur, I noticed, when I had stretched the feet widely apart, in the hollow between them a round orifice of considerable size, through which air could easily be blown into the branchial cavity, and a fine rod might even be introduced into it. The orifice opens into the branchial cavity behind a conical lobe, which stands above the third foot in place of a branchia which is wanting in *Ocypoda*. It is bounded laterally by ridges, which rise above the articulation of the foot, and to which the lower margin of the carapace is applied. Exteriorly, also, it is overarched by these ridges with the exception of a narrow fissure. This fissure is overlaid by the carapace, which exactly at this part projects further downwards than elsewhere, and in this way a complete tube is formed. Whilst in *Grapsus* the water is allowed to reach the branchiæ only from the front, I saw it in *Ocypoda* flow in also through the orifice just described.

In the position of posterior entrant orifice and the accompanying peculiarities of the third and fourth

pairs of feet, two other non-aquatic species of the same
family, which I have had the opportunity of examining,
agree with *Ocypoda*. One of these, perhaps *Gelasimus
vocans*, which lives in the mangrove swamps, and likes
to furnish the mouth of its burrow with a thick, cylin-
drical chimney of several inches in height, has the
brushes on the basal joints of the feet in question composed of ordinary hairs. The other, a smaller *Gelasimus*,
not described in Milne-Edwards' 'Natural History of
Crustacea,' which prefers drier places and is not afraid
to run about on the burning sand under the vertical
rays of the noonday sun in December, but can also
endure being in water at least for several weeks, resembles *Ocypoda* in having these brushes composed of
non-setiform, delicate hairs, indeed even more delicate and more regularly constructed than in *Ocypoda*.[4]
What may be the significance of these peculiar hairs,—
whether they only keep foreign bodies from the
branchial cavity,—whether they furnish moisture to
the air flowing past them,—or whether, as their aspect,
especially in the small *Gelasimus*, reminds one of the
olfactory filaments of the Crabs, they may also perform
similar functions,—are questions the due discussion of
which would lead us too far from our subject. Nevertheless it may be remarked that in both species, especially in *Ocypoda*, the olfactory filaments in their

---

[4] This smaller *Gelasimus* is also remarkable because the chameleon-
like change of colour exhibited by many Crabs occurs very strikingly in
it. The carapace of a male which I have now before me shone with a
dazzling white in its hinder parts five minutes since when I captured it,
at present it shows a dull gray tint at the same place.

ordinary situation are very much reduced, and when they are in the water their flagella never perform the peculiar beating movements which may be observed in other Crabs, and even in the larger *Gelasimus*; moreover, the organ of smell must probably be sought in these air-breathing Crabs, as in the air-breathing Vertebrata, at the entrance to the respiratory cavity.

So much for the facts with regard to the aerial respiration of the Crabs. It has already been indicated why Darwin's theory requires that when any peculiar arrangements exist for aerial respiration, these will be differently constructed in different families. That experience is in perfect accordance with this requirement is the more in favour of Darwin, because the schoolmen far from being able to foresee or explain such profound differences, must rather regard them as extremely surprising. If, in the nearly allied families of the Ocypodidæ and Grapsoidæ, the closest agreement prevails in all the essential conditions of their structure; if the same plan of structure is slavishly followed in every thing else, in the organs of sense, in the articulation of the limbs, in every trabecula and tuft of hairs in the complicated framework of the stomach, and in all the arrangements subserving aquatic respiration, even to the hairs of the flagella employed in cleaning the branchiæ,—why have we suddenly this exception, this complete difference, in connexion with aerial respiration?

The schoolmen will scarcely have an answer for this question, except by placing themselves on the theo-

logico-teleological stand-point which has justly fallen
into disfavour amongst us, and from which the mode
of production of an arrangement is supposed to be explained, if its " adaptation " to the animal can be demonstrated. From this point of view we might certainly
say that a widely gaping fissure which had nothing pre-
judicial in it to *Aratus Pisonii* among the foliage of
the mangrove bushes, was not suitable to the *Ocypoda*
living in sand; that in the latter, in order to prevent
the penetration of the sand, the orifice of the branchial
cavity must be placed at its lowest part, directed downwards, and concealed between broad surfaces fringed
with protective brushes of hair. It is far from the
intention of these pages to enter upon a general refu-
tation of this theory of adaptation. Indeed there is
scarcely anything essential to be added to the many
admirable remarks that have been made upon this subject since the time of Spinoza. But this may be
remarked, that I regard it as one of the most important services of the Darwinian theory that it has deprived those considerations of usefulness which are still
undeniable in the domain of life, of their mystical
supremacy. In the case before us it is sufficient to
refer to the Gelasimus of the mangrove swamps, which
shares the same conditions of life with various Grapsoidæ and yet does not agree with them, but with the
arenicolous *Ocypoda*.

CHAPTER VI.

STRUCTURE OF THE HEART IN THE EDRIOPHTHALMA.

SCARCELY less striking than the example of the air-breathing Crabs, is the behaviour of the heart in the great section Edriophthalma, which may advantageously be divided, after the example of Dana and Spence Bate, only into two orders, the Amphipoda and the Isopoda.

In the Amphipoda, to which the above-mentioned naturalists correctly refer the Caprellidæ and Cyamidæ (Latreille's *Læmodipoda*), the heart has always the same position; it extends in the form of a long tube through the six segments following the head, and has three pairs of fissures, furnished with valves, for the entrance of the blood, situated in the second, third, and fourth of these segments. It was found to be of this structure by La Valette in *Niphargus (Gammarus puteanus)*, and by Claus in *Phronima;* and I have found it to be the same in a considerable number of species belonging to the most different families.[1]

[1] The young animals in the egg, a little before their exclusion, are usually particularly convenient for the observation of the fissures in the heart; they are generally sufficiently transparent, the movements of the heart are less violent than at a later period, and they lie still even

The sole unimportant exception which I have hitherto
met with is presented by the genus *Brachyscelus*,[2] in
which the heart possesses only two pairs of fissures, as it
extends forward only into the second body-segment, and
is destitute of the pair of fissures situated in this
segment in other forms.[3]

without the pressure of a glass cover. Considering the common opinion
as to the distribution of the Amphipoda, namely, that they increase in
multiplicity towards the poles, and diminish towards the equator, it
may seem strange that I speak of a considerable number of species on
a subtropical coast. I therefore remark that in a few months and
without examining any depths inaccessible from the shore, I obtained
38 different species, of which 34 are new, which, with the previously
known species (principally described by Dana) gives 60 Brazilian
Amphipoda, whilst Kröyer in his 'Grönlands Amfipoder' was acquainted with only 28 species, including 2 Læmodipoda, from the
Arctic Seas, although these had been investigated by a far greater
number of Naturalists.

[2] According to Milne-Edwards' arrangement the females of this genus
would belong to the " Hypérines ordinaires " and the previously unknown males to the " Hypérines anormales," the distinguishing character of which, namely the curiously zigzagged inferior antennæ, is only a
sexual peculiarity of the male animals. In systematising from single
dead specimens, as to the sex, age, &c. of which nothing is known, similar
errors are unavoidable. Thus, in order to give another example of very
recent date, a celebrated Ichthyologist, Bleeker, has lately distinguished
two groups of the Cyprinodontes as follows : some, the Cyprinodontini,
have a "pinna analis non elongata," and the others, the Aplocheilini,
a "pinna analis elongata" : according to this the female of a little
fish which is very abundant here would belong to the first, and the
male to the second group. Such mistakes, as already stated, are
unavoidable by the "dry-skin" philosopher, and therefore excusable ;
but they nevertheless prove in how random a fashion the present
systematic zoology frequently goes on, without principles or sure
foundations, and how much it is in want of the infallible touchstone
for the value of the different characters, which Darwin's theory promises
to furnish.

[3] I find, in Milne-Edwards' 'Leçons sur la Physiol. et l'Anat. comp.'
iii. p. 197, the statement that, according to Frey and Leuckart, the
heart of *Caprella linearis* possesses *five* pairs of fissures. I have ex-

Considering this uniformity presented by the heart in the entire order of the Amphipoda, it cannot but seem very remarkable, that in the very next order of the Isopoda, we find it to be one of the most changeable organs.

In the cheliferous Isopods (*Tanais*) the heart resembles that of the Amphipoda in its elongated tubular form, as well as in the number and position of the fissures, but with this difference, that the two fissures of each pair do not lie directly opposite each other.

In all other Isopoda the heart is removed towards the abdomen. In the wonderfully deformed parasitic Isopods of the *Porcellanæ* (*Entoniscus porcellanæ*), the spherical heart of the female is confined to a short space of the elongated first abdominal segment, and seems to possess only a single pair of fissures. In the male of *Entoniscus Cancrorum* (n. sp.), the heart (fig. 16) is situated in the third abdominal segment. In the *Cassidinæ*, the heart

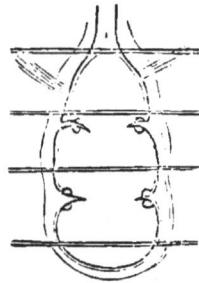

Fig. 14.[4]

(fig. 14) is likewise short and furnished with two pairs of fissures, situated in the last segment of the thorax and the first segment of the abdomen. Lastly, in a young *Anilocra*, I find the heart (fig. 15) extending through the whole length of the abdomen and furnished

amined perfectly transparent young *Caprellæ* (probably the young of *Caprella attenuata*, Dana, with which they occurred), but can only find the usual *three* pairs.

[4] Fig. 14. Heart of a young *Cassidina*.

with four (or five ?) fissures, which are not placed in
pairs but alternately to the right and left in successive
segments. In other
animals of this order,
which I have as yet
only cursorily examin-
ed, further differences
will no doubt occur.
But why, in two orders
so nearly allied to each
other, should we find
in the one such a con-
stancy, in the other
such a variability, of
the same highly im-
portant organ? From
the schoolmen we need

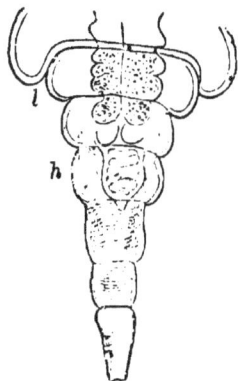

Fig. 16.⁶

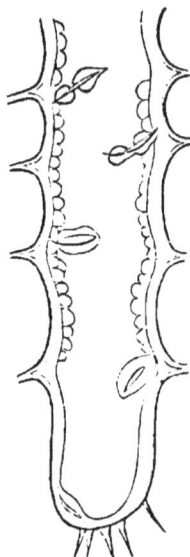

Fig. 15.⁵

expect no explanation, they will either decline the discussion of the "wherefore" as foreign to their province,
as lying beyond the boundaries of Natural History, or
seek to put down the importunate question by means of
a sounding paraphrase of the facts, abundantly sprinkled
with Greek words. As I have unfortunately forgotten
my Greek, the second way out of the difficulty is closed
to me; but as I luckily reckon myself not amongst the
incorporated masters, but, to use Baron von Liebig's
expression, amongst the "promenaders on the outskirts

⁵ Fig. 15. Heart of a young *Anilocra*.
⁶ Fig. 16. Abdomen of the male of *Entoniscus Cancrorum*. *h*. Heart.
l. Liver.

of Natural History," this affected hesitation of the schoolmen cannot dissuade me from seeking an answer, which indeed presents itself most naturally from Darwin's point of view.

As not only the *Tanaides* (which reasons elsewhere stated (*vide suprà*) justify us in regarding as particularly nearly related to the primitive Isopod) and the Amphipoda, but also the Decapod Crustacea, possess a heart with three pairs of fissures essentially in the same position; and as the same position of the heart recurs (*vide infrà*) even in the embryos of the Mantis-Shrimps (*Squilla*), in which the heart of the adult animal, and even, as I have elsewhere shown, that of the larvæ when still far from maturity, extends in the form of a long tube with numerous openings far into the abdomen, we must unhesitatingly regard the heart of the Amphipoda as the primitive form of that organ in the Edriophthalma. As, moreover, in these animals the blood flows from the respiratory organs to the heart without vessels, it is very easy to see how advantageous it must be to them to have these organs as much approximated as possible. We have reason to regard as the primitive mode of respiration, that occurring in *Tanais* (*vide suprà*). Now, where, as in the majority of the Isopoda, branchiæ were developed upon the abdomen, the position and structure of the heart underwent a change, as it approached them more nearly, but without the reproduction of a common plan for these earlier modes of structure, either because this transformation of the heart took place only after the division of the primary form into subordinate groups, or because, at least at the time of this division, the varying heart had not yet become fixed in any new form. Where, on the contrary, respiration remained with the anterior part of the body,—whether in the primitive fashion of Zoëa, as in the *Tanaides*, or by the development of branchiæ on the thorax, as in the Amphipoda,—the primitive form of the heart was inherited unchanged, because any variations which might make their appearance were rather injurious than advantageous, and disappeared again immediately.

I close this series of isolated examples with an observation which indeed only half belongs to the province of the Crustacea to which these pages ought to be confined, and which also has no further connexion with the preceding circumstances than that of being an "intelligible and intelligence-bringing fact" only from the point of view of Darwin's theory. To-day as I was opening a specimen of *Lepas anatifera* in order to compare the animal with the description in Darwin's 'Monograph on the Subclass Cirripedia,' I found in the shell of this Cirripede, a blood-red Annelide, with a short, flat body, about half an inch long and two lines in breadth, with twenty-five body-segments, and without projecting setigerous tubercles or jointed cirri. The small cephalic lobe bore four eyes and five tentacles; each body-segment had on each side at the margin a tuft of simple setæ directed obliquely upwards, and at some distance from this, upon the ventral surface, a group of thicker setæ with a strongly uncinate bidentate apex.

There was above *each* of the lateral tufts of bristles a branchia, simple on a few of the foremost segments, and then strongly arborescent to the end of the body. The animal, a female filled with ova, evidently, from these characters, belongs to the family of the Amphinomidæ; the only family the members of which, being excellent swimmers, live in the open sea.

That this animal had not strayed accidentally into the *Lepas*, but appertained to it as a regular and permanent guest, is evidenced by its considerable size in proportion to the narrow entrance of the test of the *Lepas*, by the complete absence of the iridescence which usually distinguishes the skin of free Annelides and especially of the Amphinomidæ, by the formation and position of the inferior setæ, &c. But that a worm belonging to this particular family Amphinomidæ living in the high sea, occurs as a guest in the *Lepas*, which also floats in the sea attached to wood, &c., is at once intelligible from the stand-point of the Darwinian theory, whilst the relationship of this parasite to the free-living worms of the open sea remains perfectly unintelligible under the supposition that it was independently created for dwelling in the *Lepas*.

But however favourable the examples hitherto referred to may be for Darwin, the objection may be raised against them, and that with perfect justice, that they are only isolated facts, which, when the considerations founded upon them are carried far beyond what is immediately given, may only too easily lead us from the right path, with the deceptive glimmer of an *ignis fatuus.* The higher the structure to be raised, the wider must be the assuring base of well-sifted facts.

Let us turn then to a wider field, that of the developmental history of the Crustacea, upon which science has already brought together a varied abundance of remarkable facts, which, however, have remained a barren accumulation of unmanageable raw-material, and let us see how, under Darwin's hand, these scattered stones unite to form a well-jointed structure, in which everything, bearing and being borne, finds its significant place. Under Darwin's hand! for I shall have nothing to do except just to place the building stones in the position which his theory indicates for them. "When kings build, the carters have to work."

CHAPTER VII.

DEVELOPMENTAL HISTORY OF PODOPHTHALMA.

LET us first glance over the extant facts.

Among the Stalk-eyed Crustacea (*Podophthalma*) we know only a very few species which quit the egg in the form of their parents, with the full number of well-jointed appendages to the body. This is the case according to Rathke[1] in the European fresh-water Crayfish, and according to Westwood in a West Indian Land Crab (*Gecarcinus*). Both exceptions therefore belong to the small number of Stalk-eyed Crustacea which live in fresh water or on the land, as indeed in many other cases fresh-water and terrestrial animals undergo no transformations, whilst their allies in the sea have a metamorphosis to undergo. I may refer to the Earth-worms and Leeches among the Annelida, which chiefly belong to the land and to fresh water,—to the *Planariæ* of the fresh waters and the *Tetrastemma* of the sparingly saline Baltic among the Turbellaria,—to the Pulmonate Gasteropoda, and to the Branchiferous Gasteropoda of the fresh waters, the young of which (according to

[1] Authorities are cited only for facts which I have had no opportunity of confirming.

Troschel's 'Handb. der Zoologie') have no ciliated buccal lobes, although such organs are possessed by the very similar Periwinkles (*Littorina*).

All the marine forms of this section appear to be subject to a more or less considerable metamorphosis. This appears to be only inconsiderable in the common Lobster, the young of which, according to Van Beneden, are distinguished from the adult animal, by having their feet furnished, like those of *Mysis*, with a swimming branch projecting freely outwards. From a figure given by Couch the appendages of the abdomen and tail also appear to be wanting.

Far more profound is the difference of the youngest brood from the sexually mature animal in by far the greater majority of the Podophthalma, which quit the egg in the form of *Zoëa*. This young form occurs, so far as our present observations go, in all the Crabs, with the sole exception of the single species investigated by Westwood. I say *species*, and not *genus*, for in the same genus, *Gecarcinus*, Vaughan Thompson found Zoëa-brood,[2] which is also met with in other terrestrial crabs (*Ocypoda*, *Gelasimus*, &c.). All the Anomura

---

[2] Bell ('Brit. Stalk-eyed Crust.' p. xlv.) considers himself justified in "eliminating" Thompson's observation at once, because he could only have examined ovigerous females preserved in alcohol. But any one who had paid so much attention as Thompson to the development of these animals, must have been well able to decide with certainty upon eggs, if not too far from maturity or badly preserved, whether a Zoëa would be produced from them. Moreover, the mode of life of the Land-Crabs is in favour of Thompson. "Once in the year," says Troschel's 'Handbuch der Zoologie,' "they migrate in great crowds to the sea in order to deposit their eggs, and afterwards return much exhausted

seem likewise to commence their lives as Zoëæ:
witness the *Porcellanæ*, the Tatuira (*Hippa emerita*) and
the Hermit Crabs.   Among the Macrura we are acquainted with the same earliest form principally in
several Shrimps and Prawns, such as *Crangon* (Du
Cane), *Caridina* (Joly), *Hippolyte, Palæmon, Alpheus,*
&c.   Lastly, it is not improbable, that the youngest
brood of the Mantis-Shrimps (*Squilla*) is also in the
same case.

The most important peculiarities which distinguish
this Zoëa-brood from the adult animal, are as follows :—

The middle-body with its appendages, those five pairs
of feet to which these animals owe their name of Decapoda, is either entirely wanting, or scarcely indicated ;
the abdomen and tail are destitute of appendages, and
the latter consists of a single piece.   The mandibles, as in
the Insecta, have no palpi.   The maxillipedes, of which
the third pair is often still wanting, are not yet brought
into the service of the mouth, but appear in the form
of biramose natatory feet.   Branchiæ are wanting, or
where their first rudiments may be detected as small
verruciform prominences, these are dense cell-masses,
through which the blood does not yet flow, and which
therefore have nothing to do with respiration.   An interchange of the gases of the water and blood may occur
all over the thin-skinned surface of the body ; but the

towards their dwelling places, which are reached only by a few." For
what purpose would be these destructive migrations in species whose
young quit the egg and the mother as terrestrial animals ?

lateral parts of the carapace may unhesitatingly be indicated as the chief seat of respiration. They consist,
exactly as described by Leydig in the *Daphniæ*, of an
outer and inner lamina, the space between which is
traversed by numerous transverse partitions dilated at
their ends; the spaces between these partitions are
penetrated by a more abundant flow of blood than
occurs anywhere else in the body of the Zoëa. To this

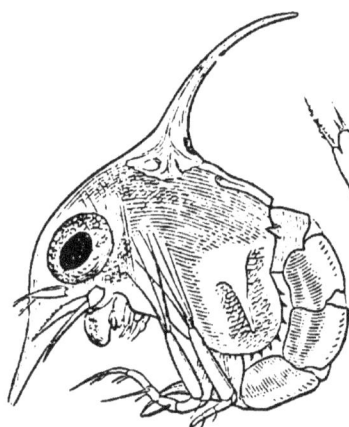

Fig. 17.³   Fig. 18.⁴

may be added that a constant current of fresh water
passes beneath the carapace in a direction from behind
forwards, maintained as in the adult animal, by a foliaceous or linguiform appendage of the second pair of
maxillæ (fig. 18). The addition of fine coloured par-
ticles to the water allows this current of water to be
easily detected even in small Zoëæ.

³ Fig. 17. Zoëa of a Marsh Crab (*Cyclograpsus*?), magn. 45 diam.
⁴ Fig. 18. Maxilla of the second pair in the same species, magn.
180 diam.

The Zoëæ of the Crabs (fig. 17) are usually distinguished by long, spiniform processes of the carapace.
One of these projects upwards from the middle of the
back, a second downwards from the forehead, and frequently there is a shorter one on each side near the
posterior inferior angles of the carapace.   All these
processes are, however, wanting in *Maia* according to
Couch, and in *Eurynome* according to Kinahan; and
in a third species of the same group of the *Oxyrhynchi*
(belonging or nearly allied to the genus *Achæus*) I also
find only an inconsiderable dorsal spine, whilst the forehead and sides are unarmed.   This is another example
warning us to be cautious in deductions from analogy.
Nothing seemed more probable than to refer back the
beak-like formation of the forehead in the Oxyrhynchi
to the frontal process of the Zoëa, and now it appears
that the young of the Oxyrhynchi are really quite
destitute of any such process.   The following are more
important peculiarities of the Zoëæ of the Crabs,
although less striking than these processes of the carapace which, in combination with the large eyes, often
give them so singular an appearance:—the anterior
(inner) antennæ are simple, not jointed, and furnished
at the extremity with from two to three olfactory filaments; the posterior (outer) antennæ frequently run
out into a remarkably long spine-like process (" styli-
form process," Spence Bate), and bear, on the outside,
an appendage, which is sometimes very minute ("squami-
form process" of Spence Bate), corresponding with the

E 2

antennal scale of the Prawns,[5] and the first rudiment of the future flagellum is often already recognisable. Of natatory feet (afterwards maxillipeds) only two pairs are present; the third (not, as Spence Bate thinks, the first) is entirely wanting, or, like the five following pairs of feet, present only as a minute bud. The tail, of very variable form, always bears *three* pairs of setæ at its hinder margin. The Zoëæ of the Crabs usually

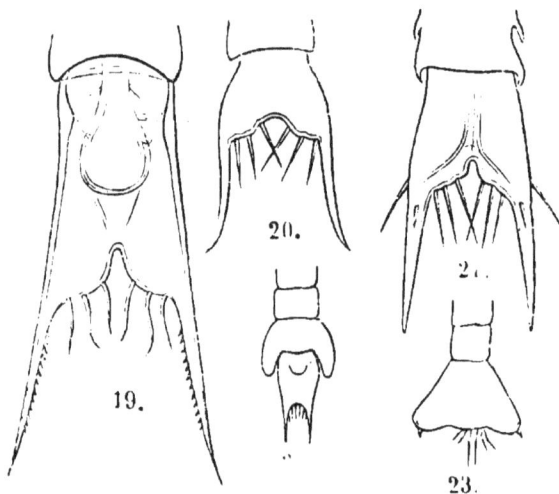

Figs. 19—23.[6]

maintain themselves in the water in such a manner that the dorsal spine stands upwards, the abdomen is bent forwards, the inner branch of the natatory feet is directed forwards, and the outer one outwards and upwards.

---

[5] In a memoir on the metamorphoses of the *Porcellanæ* I have erroneously described this appendage as the "flagellum."

[6] Tails of the Zoëæ of various Crabs. Fig. 19. *Pinnotheres*. Fig. 20. *Sesarma*. Fig. 21. *Xantho*. Figs. 22 and 23 of unknown origin.

It is further to be remarked that the Zoëæ of the crabs, as also of the *Porcellanæ*, of the Tatuira and of the Shrimps and Prawns, are enveloped, on escaping from the egg, by a membrane veiling the spinous processes of the carapace, the setæ of the feet, and the antennæ, and that they cast this in a few hours. In *Achæus* I have observed that the tail of this earliest larval skin resembles that of the larvæ of Shrimps and Prawns, and the same appears to be the case in *Maia* (see Bell, ' Brit. Stalk-eyed Crust.,' p. 44).

Widely as they seem to differ from them at the first glance, the Zoëæ of the *Porcellanæ* (fig. 24) approach those of the true Crabs very closely. The antennæ, organs of the mouth, and natatory feet, exhibit the same structure. But the tail bears *five* pairs of setæ, and the dorsal spine is wanting, whilst, on the contrary, the frontal process and the lateral spines are of extraordinary length, and directed straight forward and backward.

The Zoëa of the Tatuira (fig. 25) also appears to differ but little from

---

[7] Fig. 24. Zoëa of *Porcellana stellicola*, F. Müll. Magn. 15 diam.

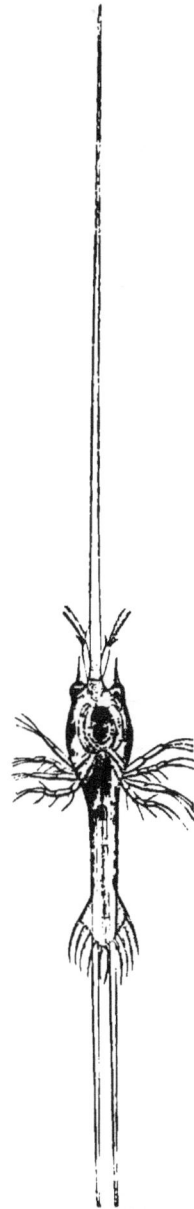

Fig. 24.[7]

those of the true Crabs, which it likewise resembles
in its mode of locomotion. The carapace
possesses only a short, broad frontal process;
the posterior margin of the tail is edged
with numerous short setæ.

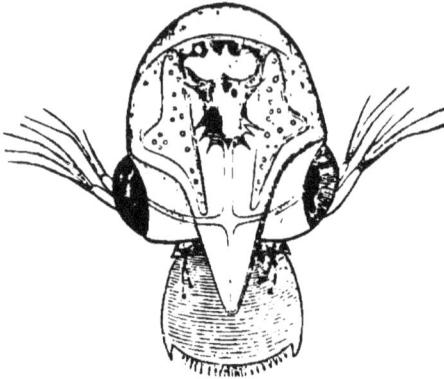

Fig. 25.[8]

The Zoëa of the
Hermit Crabs (fig. 26)
possesses the simple
inner antennæ of the
Zoëa of the true Crabs;
the outer antennæ bear
upon the outside on a
short stalk a lamella of
considerable size ana-
logous to the scale of
the antennæ of the
Prawns; on the inside,
a short, spine-like pro-
cess; and between the
two the flagellum, still
short, but already fur-
nished with two apical

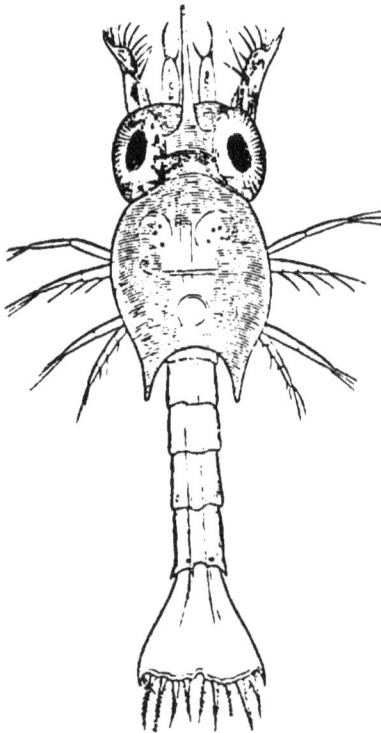

Fig. 26.[9]

[8] Fig. 25. Zoëa of the Tatuira (*Hippa emerita*), magn.
45 diam.

[9] Fig. 26. Zoëa of a small
Hermit Crab, magn. 45 diam.

setæ. As in the Crabs, there are only two pairs of
well-developed natatory feet (maxillipedes), but the
third pair is also present in the form of a two-jointed
stump of considerable size, although still destitute of
setæ. The tail bears five pairs of setæ. The little
animal usually holds itself extended straight in the
water, with the head directed downwards.

This is also the position in which we usually see the
Zoëæ of the Shrimps and Prawns (fig. 27), which agree
in their general appearance with those of the Hermit
Crabs. Between the large compound eyes there is in
them a small median eye. The inner antennæ bear, at
the end of a basal joint sometimes of considerable length,
on the inside a plumose seta, which also occurs in the
Hermit Crabs, and on the outside a short terminal joint
with one or more olfactory filaments. The outer
antennæ exhibit a well-developed and sometimes dis-
tinctly articulated scale, and within this usually a spiniform process; the flagellum appears generally to be still
wanting. The third pair of maxillipedes seems to be
always present, at least in the form of considerable
rudiments. The spatuliform caudal lamina bears from
five to six pairs of setæ on its hinder margin.

The development of the Zoëa-brood to the sexually
mature animal was traced by Spence Bate in *Carcinus
mænas*. He proved that the metamorphosis is a perfectly gradual one, and that no sharply separated stages
of development, like the caterpillar and pupa of the
Lepidoptera, could be defined in it. Unfortunately we
possess only this single complete series of observations,

and its results cannot be regarded at once as universally applicable; thus the young Hermit Crabs retain the general aspect and mode of locomotion of Zoëæ, whilst the rudiments of the thoracic and abdominal feet are growing, and then, when these come into action, appear at once in a perfectly new form, which differs from that of the adult animal chiefly by the complete symmetry of the body and by the presence of four pairs of well-developed natatory feet on the abdomen.[10]

The development of the Palinuridæ seems to be very peculiar. Claus found in the ova of the Spiny Lobster (*Palinurus*), embryos with a completely segmented body, but wanting the appendages of the tail, abdomen, and last two segments of the middle-body; they possess a single median and considerably compound eye; the anterior antennæ are simple, the posterior furnished with a small

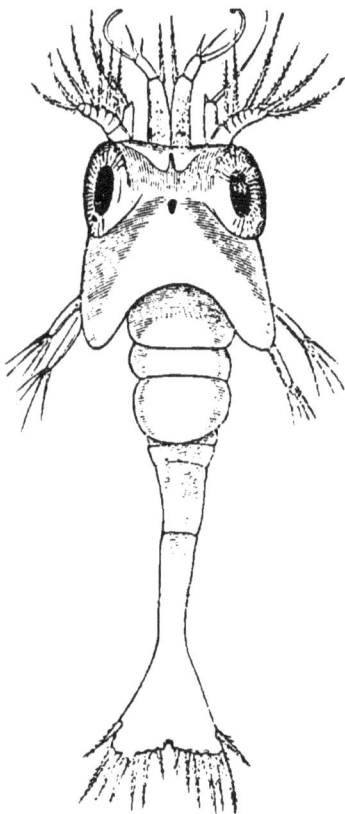

Fig. 27.[11]

[10] *Glaucothoë Peronii*, M.-Edw., may be a young and still symmetrical *Pagurus* of this kind.

[11] Fig. 27. Zoëa of a *Palæmon* residing upon *Rhizostoma cruciatum*, Less., magn. 45 diam.

secondary branch; the mandibles have no palpi; the maxillipedes of the third pair, like the two following pairs of feet, are divided into two branches of nearly equal length; whilst the last of the existing pairs of feet and the second pair of maxillipedes bear only an inconsiderable secondary branch. Coste, as is well known, asserts that he has bred young *Phyllosomata* from the ova of this lobster—a statement that requires further proof, especially as the more recent investigations of Claus upon *Phyllosoma* by no means appear to be in its favour.

The large compound eyes, which usually soon become moveable, and sometimes stand upon long stalks even in the earliest period, as well as the carapace, which covers the entire fore-body, indicate at once that the position of the larvæ hitherto considered, notwithstanding all their differences, is under the Podophthalma. But not a single characteristic of this section is retained by the brood of some Prawns belonging to the genus *Penëus* or in its vicinity. These quit the egg with an unsegmented ovate body, a median frontal eye, and three pairs of natatory feet, of which the anterior are simple, and the other two biramose—in fact, in the larval form, so common among the lower Crustacea, to which O. F. Müller gave the name of *Nauplius*. No trace of a carapace! no trace of the paired eyes! no trace of masticating organs near the mouth which is overarched by a helmet-like hood!

In the case of one of these species the intermediate forms which lead from the Nauplius to the Prawn, have been discovered in a nearly continuous series.

The youngest Nauplius (fig. 28) is immediately followed by forms in which a fold of skin runs across the
back behind the third pair of feet, and four pairs of
stout processes (rudiments of new limbs) sprout forth
on the ventral surface. Within the third pair of feet,
powerful mandibles are developed.

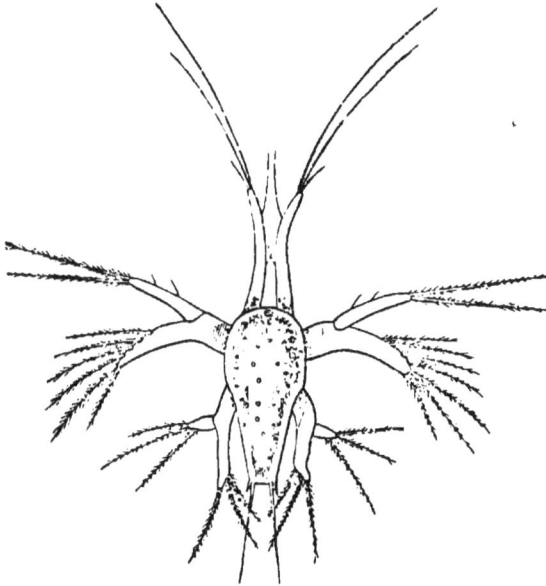

Fig. 28.[12]

In a subsequent moult the new limbs (maxillæ, and
anterior and intermediate maxillipedes) come into action, and in this way the Nauplius becomes a Zoëa
(fig. 29), agreeing perfectly with the Zoëa of the Crabs
in the number of the appendages of the body, although
very different in form and mode of locomotion and even
in many particulars of internal structure. The chief

[12] Fig. 28. Nauplius of a Prawn, magn. 45 diam.

organs of motion are still the two anterior pairs of feet, which are slender and furnished with long setæ; the third pair of feet loses its branches, and becomes converted into mandibles destitute of palpi.   The labrum acquires a spine directed forward and of considerable size, which occurs in all the Zoëæ of allied species.   The biramose maxillipedes appear to assist but slightly in locomotion. The forked tail reminds us rather of the forms occurring in the lower Crustacea, especially the Copepoda, than of the spatuliform caudal plate which characterises the Zoëæ of *Alpheus, Palæmon, Hippolyte,* and other Prawns,

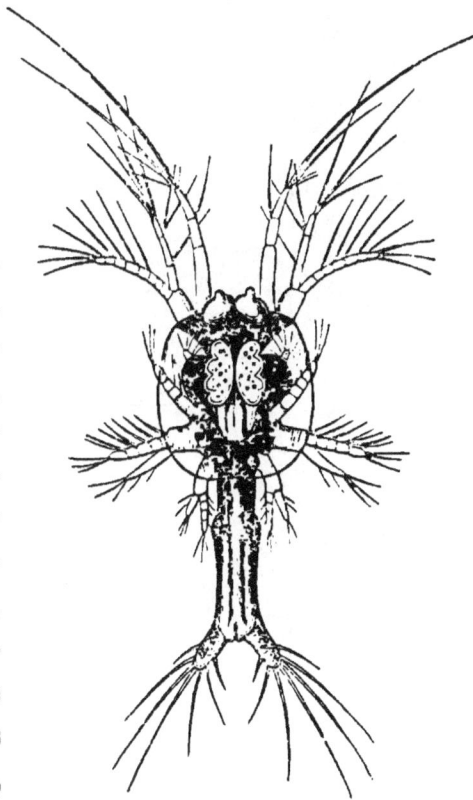

Fig. 29.[13]

of the Hermit Crabs, the Tatuira and the *Porcellanæ.* The heart possesses only one pair of fissures, and has no muscles traversing its interior like trabeculæ, whilst in other Zoëæ two pairs of fissures and an interior appa-

[13] Fig. 29. Young Zoëa of the same Prawn, magn. 45 diam.

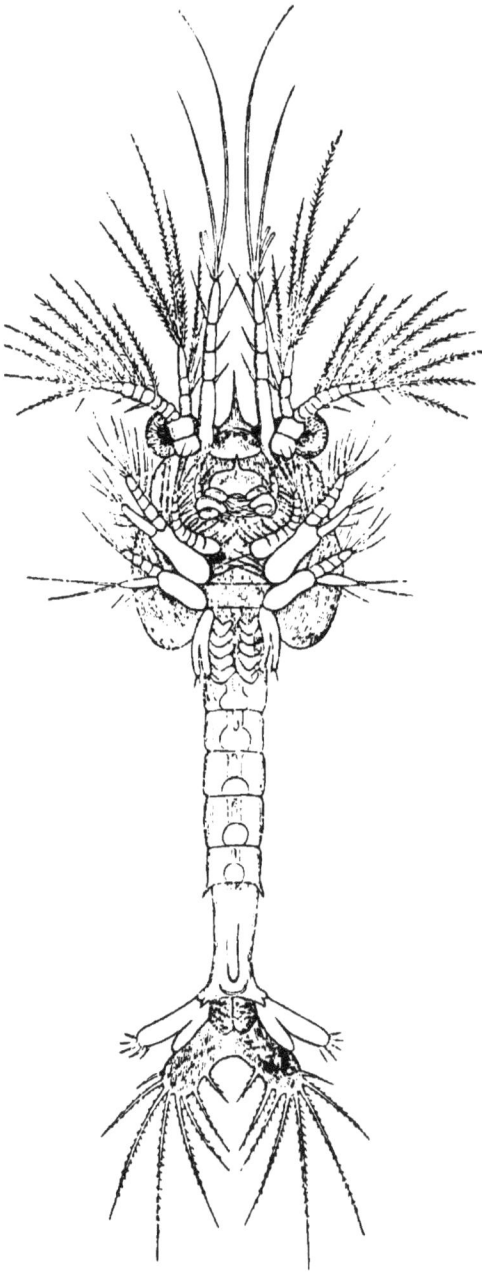

Fig. 30.[14]

ratus of trabeculæ
are always distinct-
ly recognisable.

During this Zoëal
period the paired
eyes, the segments
of the middle-body
and abdomen, the
posterior maxilli-
pedes, the lateral
caudal appendages
and the stump-like
rudiments of the
feet of the middle-
body are formed
(fig. 30). The caudal
appendages sprout
forth like other
limbs freely on the
ventral surface,
whilst in other
Prawns, the *Porcel-
lanæ*, &c., they are
produced in the in-
terior of the spatuli-
form caudal plate.

As the feet of the
middle-body come

[14] Fig. 30. Older Zoëa
of the same Prawn, magn.
45 diam.

into action, simultaneously with other profound changes, the Zoëa passes into the *Mysis*- or Schizopod-form (fig.31). The antennæ cease to serve for locomotion, their place is taken by the thoracic feet, furnished with long setæ, and by the long abdomen which just before was laboriously dragged along as a useless burden, but now, with its powerful muscles, jerks the animal through the water in a series of lively jumps. The anterior antennæ have lost their long setæ, and by the side of the last (fourth) joint, endowed with olfactory filaments, there appears a second branch, which is at first of a single joint. The previously multi-articulate outer branch of the posterior antennæ has become a simple lamella, the antennal scale of the Prawn; beside this appears the stump-like rudiment of the flagellum, probably as a new formation, the inner branch disappearing entirely. The five new pairs of feet are biramose,

Fig. 31.[15]

the inner branch short and simple, the outer one longer, annulated at the end, furnished with long setæ, and

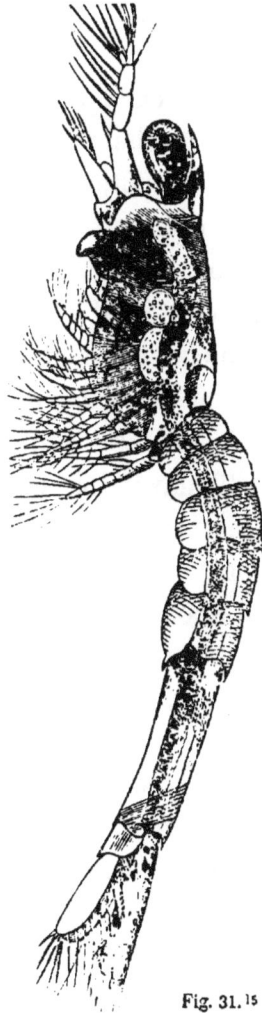

[15] Fig. 31. *Mysis*-form of the same Prawn, magn. 45 diam.

kept, as in *Mysis*, in constant whirling motion. The
heart acquires new fissures, and interior muscular
trabeculæ.

During the *Mysis*-period, the auditory organs in the
basal joint of the anterior antennæ are formed; the
inner branches of the first three pairs of feet are developed into chelæ and the two hinder pairs into ambulatory feet; palpi sprout from the mandibles, branchiæ
on the thorax, and natatory feet on the abdomen. The
spine on the labrum becomes reduced in size. In this
way the animal gradually approaches the Prawn-form,
in which the median eye has become indistinct, the
spine of the labrum, and the outer branches of the
cheliferous and ambulatory feet have been lost, the
mandibular palpi and the abdominal feet have acquired
distinct joints and setæ, and the branchiæ come into
action.

In another Prawn, the various larval states of which
may be easily recognised as belonging to the same
series by the presence of a dark-yellow, sharply-defined
spot surrounding the median eye, the youngest Zoëa
(fig. 32), probably produced from the Nauplius, agrees
in all essential particulars with the species just described; its further development is, however, very different, especially in that neither the feet of the middle,
nor those of the hind-body are formed simultaneously,
and that a stage of development comparable to *Mysis* in
the number and structure of the limbs does not occur.

Traces of the outer maxillipedes make their appearance betimes. Then feet appear upon four segments

of the middle-body, and these are biramose on the three
anterior segments, and simple, the inner branch being
deficient, on the fourth
segment. On the inner
branches the chelæ are
developed ; the outer
branches are lost before
an inner branch has
made its appearance on
the fourth segment (fig.
32). The latter again
becomes destitute of ap-
pendages, so that in this
case at an early period
four, and at a later only
three, segments of the
middle-body bear limbs.
The fifth segment is still
entirely wanting, whilst
all the abdominal seg-
ments have also acquired

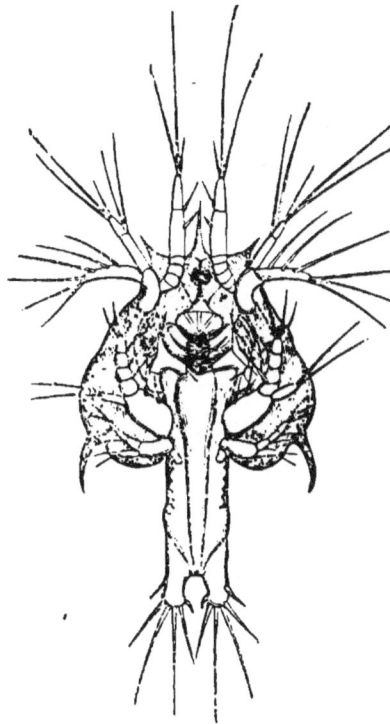

Fig. 32.[16]

limbs, and this one after the other, from before backwards. The adult animal, as shown by the three pairs
of chelæ, will certainly be very nearly allied to the
preceding species.[17]

---

[16] Fig. 32. Youngest (observed) Zoëa of another Prawn. The
minute buds of the third pair of maxillipedes are visible. The formation of the abdominal segments has commenced. Paired eyes still
wanting. Magn. 45 diam.

[17] The oldest observed larvæ (see fig. 33) are characterised by the
extraordinary length of the flagella of the outer antennæ, and in this

The youngest larva of the Schizopod genus *Euphausia* observed by Claus, stands very near the youngest Zoëa of our Prawns; but whilst its anterior antennæ are already biramose, and it therefore appears to be more advanced, it still wants the middle maxillipedes. In it also Claus found the heart furnished with only a single pair of fissures. Do not Nauplius-like states in this case also precede the Zoëa?

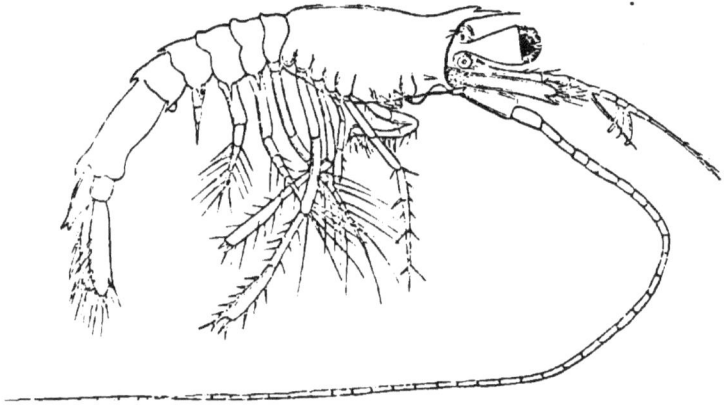

Fig. 33. [8]

The developmental history of *Mysis*, the near relationship of which with the Shrimps and Prawns has recently again been generally recognised, has been

respect resemble the larva of *Sergestes* found by Claus near Messina (Zeitschr. für Wiss. Zool. Bd. xiii. Taf. 27, fig. 14). This unusual length of the antennæ leads to the supposition that they belong to our commonest Prawn, which is very frequently eaten, and is most nearly allied to *Penëus setiferus* of Florida. Claus's *Acanthosoma* (*l. c.* fig. 13) is like the younger *Mysis*-form of the larva figured by me in the 'Archiv für Naturgeschichte,' 1836, Taf. 2, fig. 18, and which I am inclined to refer to *Sicyonia carinata*.

[18] Fig. 33. Older larva produced from the Zoëa represented in fig. 32. The last segment and the last two pairs of feet of the middle-body are wanting. Magn. 20 diam.

described in detail by Van Beneden. So far as I have tested them I can only confirm his statements. The development of the embryo commences with the formation of the tail! This makes its appearance as a simple lobe, the dorsal surface of which is turned towards and closely applied to that of the embryo. (The young of other Stalk-eyed Crustacea are, as is well known, bent in the egg in such a manner that the ventral surfaces of the anterior and posterior halves of the body are turned towards each other,—in these, therefore, the dorsal, and in Mysis the ventral surface appears convex.) The tail soon acquires the furcate form with which we made acquaintance in the last Prawn-Zoëa described. Then two pairs of thick ensiform appendages make their appearance at the opposite end of the body, and behind these a pair of tubercles which are easily overlooked. These are the antennæ and mandibles. The egg-membrane now bursts, before any internal organ, or even any tissue, except the cells of the cutaneous layer, is formed. The young animal might be called a Nauplius; but essentially there is nothing but a rough copy of a Nauplius-skin, almost like a new egg-membrane, within which the *Mysis* is developed. The ten pairs of appendages of the fore- (maxillæ, maxillipedes) and middle-body make their appearance simultaneously, as do the five pairs of abdominal feet at a later period. Soon after the young *Mysis* casts the Nauplius-envelope it quits the brood-pouch of the mother.[19]

[19] Van Beneden, who regards the eye-peduncles as limbs, cannot however avoid remarking upon *Mysis:* "Ce pédicule n'apparaît

F

For some time, owing to an undue importance being ascribed to the want of a particular branchial cavity, *Mysis, Leucifer,* and *Phyllosoma* were referred to the Stomapoda, which are now again limited, as originally by Latreille, to the Mantis-shrimps (*Squilla*), the Glass-shrimps (*Erichthus*) and their nearest allies. Of the developmental history of these we have hitherto been acquainted with only isolated fragments. The tracing of the development in the egg is rendered difficult by the circumstance, that the Mantis-shrimps do not, like the Decapoda, carry their spawn about with them, but deposit it in the subterranean passages inhabited by them in the form of thin, round, yellow plates. The spawn is consequently exceedingly difficult to procure, and unfortunately it becomes spoilt in a day when it is removed from its natural hatching place, whilst on the contrary the progress of development may be followed for weeks together in the eggs of a single Crab kept in confinement. The eggs of *Squilla,* like those removed from the body of the Crab, die because they are deprived of the rapid stream of fresh water which the mother drives through her hole for the purpose of her own respiration.

The accompanying representation of the embryo of *Squilla* shows that it possesses a long, segmented abdomen without appendages, a bilobate tail, six pairs of limbs, and a short heart; the latter only pulsates weakly and slowly. If it acquires more limbs before

aucurement comme les autres appendices, et paraît avoir une autre valeur morphologique."

exclusion, the youngest larva must stand on the same level as the youngest larva of *Euphausia* observed by Claus.

Of the two larval forms at present known which are with certainty to be ascribed, if not to *Squilla*, at least to a Stomapod, I pass over the younger one [21] as its limbs cannot be positively interpreted, and will only mention that in it

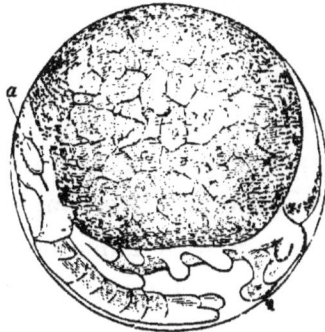

Fig. 31.[20]

the last three abdominal segments are still destitute of appendages. The older larva (fig. 35), which resembles the mature *Squilla* especially in the structure of the great raptorial feet and of the preceding pair, still wants the six pairs of feet following the raptorial feet. The cor-responding body-

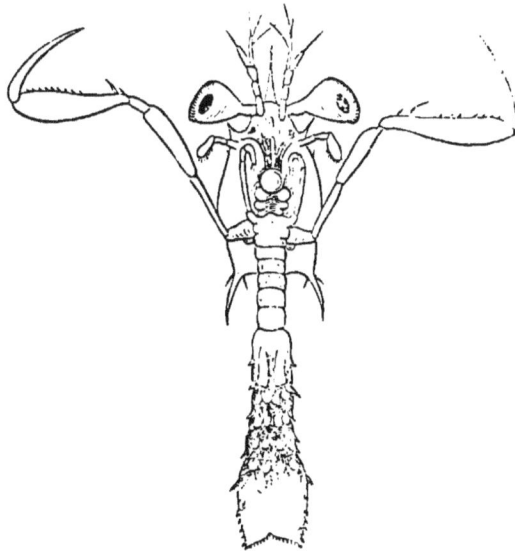

Fig 35. [22]

segments are already well developed, an unpaired eye

[20] Fig. 34. Embryo of a Squilla, magn. 45 diam. *a.* heart.
[21] 'Archiv für Naturgeschichte,' 1863.   Taf. 1.
[22] Fig. 35. Older larva (Zoëa) of a Stomapod, magn. 15 diam.

is still present, the anterior antennæ are already biramose, whilst the flagellum is wanting in the posterior, and
the mandibles are destitute of palpi; the four anterior
abdominal segments bear biramose natatory feet, without branchiæ; the fifth abdominal segment has no
appendages, and this is also the case with the tail, which
still appears as a simple lamina, fringed on the hinder
margin with numerous short teeth. It is evident that
the larva stands essentially in the grade of Zoëa.

CHAPTER VIII.

DEVELOPMENTAL HISTORY OF EDRIOPHTHALMA.

LESS varied than that of the Stalk-eyed Crustacea is
the mode of development of the Isopoda and Amphipoda, which Leach united in the section Edriophthalma,
or Crustacea with sessile eyes.

The Rock-Slaters (*Ligia*) may serve as an example
of the development of the
Isopoda. In these, as in
Mysis, the caudal portion
of the embryo is bent
not downwards, but up-
wards; as in *Mysis* also,
a larval membrane is
first of all formed, within

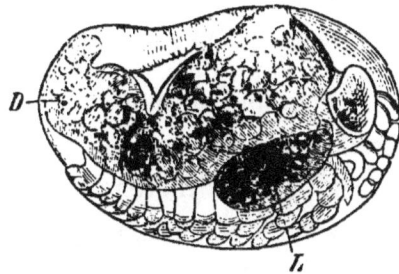

Fig. 36. [1]

which the Slater is developed. In *Mysis* this first larval
skin may be compared to a Nauplius; in *Ligia* it appears
like a maggot quite destitute of appendages, but produced into a long simple tail (fig. 37). The egg-membrane is retained longer than in *Mysis;* it bursts only
when the limbs of the young Slater are already partially developed in their full number. The dorsal sur-

[1] Fig. 36. Embryo of *Ligia* in the egg, magn. 15 diam. *D.* yelk;
L. liver.

face of the Slater is united to the larval skin a little behind the head. At this point, when the union has been dissolved a little before the change of skin, there is a foliaceous appendage, which exists only for a short time, and disappears before the young Slater quits the brood-pouch of the mother.

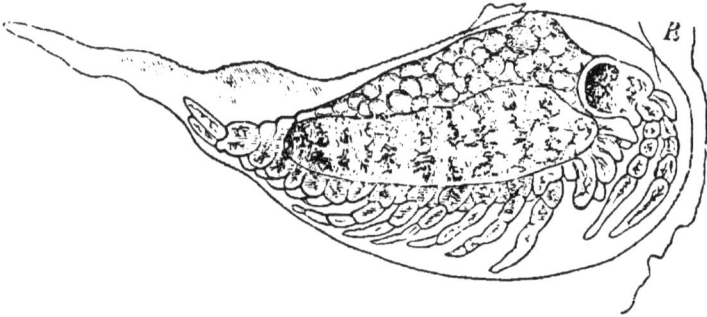

Fig. 37.[2]

The young animal, when it begins to take care of itself, resembles the old ones in almost all parts, except one important difference; it possesses only six, instead of seven pairs of ambulatory feet; and the last segment of the middle-body is but slightly developed and destitute of appendages. It need hardly be mentioned that the sexual peculiarities are not yet developed, and that in the males the handlike enlargements of the anterior ambulatory feet and the copulatory appendages are still deficient.

[2] Fig. 37. Maggot-like larva of *Ligia*, magn. 15 diam. *R.* remains of the egg-membrane. We see on the lower surface, from before backwards:—the anterior and posterior antennæ, the mandibles, the anterior and posterior maxillæ, maxillipedes, six ambulatory feet, the last segment of the middle-body destitute of appendages, five abdominal feet, and the caudal feet.

To the question, how far the development of *Ligia* is repeated in the other Isopoda, I can only give an unsatisfactory answer. The curvature of the embryo upwards instead of downwards was met with by me as well as by Rathke in *Idothea*, and likewise in *Cassidina*, *Philoscia*, *Tanais*, and the Bopyridæ,—indeed, I failed to find it in none of the Isopoda examined for this purpose. In *Cassidina* also the first larval skin without appendages is easily detected; it is destitute of the long tail, but is strongly bent in the egg, as in *Ligia*, and consequently cannot be mistaken for an "inner egg-membrane." This, however, might happen in *Philoscia*, in which the larval skin is closely applied to the egg-membrane (fig. 38), and is only to be explained as the larval skin by a reference to *Ligia* and *Cassidina*.

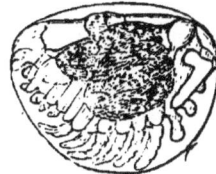

Fig. 38.[3]

The foliaceous appendage on the back has long been known in the young of the common Water Slater (*Asellus*).[4] That the last pair of feet of the thorax

---

[3] Fig. 38. Embryo of a *Philoscia* in the egg, magn. 25 diam.

[4] Leydig has compared this foliaceous appendage of the Water Slaters with the "green gland" or "shell-gland" of other crustacea, assuming that the green gland has no efferent duct and appealing to the fact that the two organs occur "in the same place." This interpretation is by no means a happy one. In the first place we may easily ascertain in *Leucifer*, as was also found to be the case by Claus, that the "green gland" really opens at the end of the process described by Milne-Edwards as a "tubercule auditif" and by Spence Bate as an "olfactory denticle." And, secondly, the position is about as different as it can well be. In the one case a paired gland, opening at the base of the posterior antennæ, and therefore on the lower surface of the *second* segment; in the other an unpaired structure rising

is wanting in the young of the Wood-lice (*Porcellionides,*
M.-Edw.) and Fish-lice (*Cymothoadiens,* M.-Edw.) has
already been noticed by Milne-Edwards. This applies
also to the Box-Slaters (*Idothea*), to the viviparous Globe-
Slaters (*Sphæroma*) and Shield-Slaters (*Cassidina*), to
the Bopyridæ (*Bopyrus, Entoniscus, Cryptoniscus,* n. g.),
and to the Cheliferous Slaters (*Tanais*), and therefore
probably to the great majority of the Isopoda. All the
other limbs are usually well developed in the young
Isopoda. In *Tanais* alone, all the abdominal feet are
wanting (but not those of the tail); they are developed
simultaneously with the last pair of feet of the thorax.

The last pair of feet on the middle-body of the larva,
consequently the penultimate pair in the adult animal,
is almost always similar in structure to the preceding
pair. A remarkable excep-
tion is, however, presented
in this respect by *Cryptoniscus* and *Entoniscus,* — remarkable as a confirmation
of Darwin's proposition that
" parts developed in an unusual manner are very variable," for in the peculiarly-formed pair of feet there
exists the greatest possible difference between the three
species hitherto observed. In *Cryptoniscus* (fig. 39)
this last foot is thin and rod-like ; in *Entoniscus Can-*

Fig. 39. [5]   Fig. 40. [6]

in the median line of the back *behind the seventh segment,* (" behind the
boundary line of the first thoracic segment," Leydig).

[5] Fig. 39. Embryo of *Cryptoniscus planarioïdes,* magn. 90 diam.

[6] Fig. 40. Last foot of the middle-body of the larva of *Entoniscus
Porcellanæ,* magn. 180 diam.

crorum remarkably long and furnished with a strongly
thickened hand and a peculiarly constructed chela ; in
Entoniscus Porcellanæ very short, imperfectly jointed,
and with a large ovate terminal joint (fig. 40).

Some Isopods undergo a considerable change immediately before the attainment of sexual maturity. This
is the case with the males of *Tanais* which have already
been noticed, and, according to Hesse, with the *Pra-
nizæ*, in which both sexes are said to pass into the form
known as *Anceus*. But Spence Bate, a careful observer,
states that he has seen females of the form of *Praniza*
laden with eggs far advanced in their development.

In this order we meet for the first time with an
extensive retrograde metamorphosis as a consequence
of a parasitic mode of life. Even in some Fish-lice
(*Cymothoa*) the young are lively swimmers, and the
adults stiff, stupid, heavy fellows, whose short clinging
feet are capable of but little movement. In the Bopyridæ (*Bopyrus, Phryxus, Kepone,* &c., which might
have been conveniently left in a single genus), which
are parasitic on Crabs, Lobsters, &c., taking up their
abode chiefly in the branchial cavity, the adult females
are usually quite destitute of eyes; the antennæ are
rudimentary ; the broad body is frequently unsymme-
trically developed in consequence of the confined
space ; its segments are more or less amalgamated with
each other; the feet are stunted, and the appendages
of the abdomen transformed from natatory feet with
long setæ into foliaceous or tongue-shaped and sometimes ramified branchiæ. In the dwarfish males the

eyes, antennæ, and feet, are usually better preserved
than in the females; but on the other hand all the
appendages of the abdomen have not unfrequently disappeared, and sometimes every trace of segmentation.
In the females of *Entoniscus*, which are found in the
body-cavity of Crabs and *Porcellanæ*, the eyes, antennæ,
and buccal organs, the segmentation of the vermiform

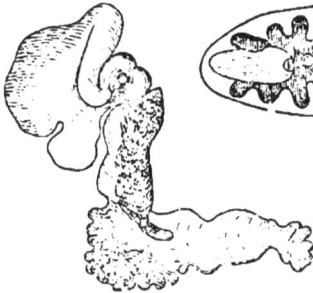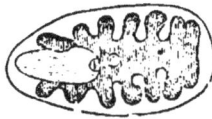

body, and in one spe-
cies (fig. 41) the whole
of the limbs, disappear
almost without leaving
a trace; and *Cryptonis-
cus planarioïdes* would
almost be regarded as a
Flatworm rather than
an Isopod, if its eggs

Fig. 41. [7]   Fig. 42. [8]

and young did not betray its Crustacean nature. Among
the males of these various Bopyridæ, that of *Entoniscus
Porcellanæ* occupies the lowest place; it is confined all

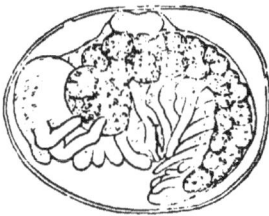

its life to six pairs of feet, which
are reduced to shapeless rounded
lumps.

The Amphipoda are distinguish-
able from the Isopoda at an early
period in the egg by the different
position of the embryo, the hinder

Fig. 43. [9]

extremity of which is bent downwards. In all the ani-

---

[7] Fig. 41. *Entoniscus Cancrorum*, female, magn. 3 times.
[8] Fig. 42. *Cryptoniscus planarioïdes*, female, magn. 3 times.
[9] Fig. 43. Embryo of a *Corophium*, magn. 90 diam.

mals of this order which have been examined for it,[10]
a peculiar structure makes its appearance very early
on the anterior part of the back, by which the embryo is
attached to the " inner egg-membrane," and which has
been called the " micropylar apparatus," but improperly
as it seems to me.[11]   It will remind us of the union of the
young Isopoda with the larval membrane and of the unpaired " adherent organ " on the nape of the Cladocera,
which is remarkably developed in *Evadne* and persists
throughout life ; but in *Daphnia pulex*, according to
Leydig, although present in the young animals, disappears without leaving a trace in the adults.

The young animal, whilst still in the egg, acquires the
full number of its segments and limbs.   In cases where
segments are amalgamated together, such as the last two
segments of the thorax in *Dulichia*, the last abdominal
segments and the tail in *Gammarus ambulans* and *Coro-*

---

[10] In the genera *Orchestoidea, Orchestia, Allorchestes, Montagua,
Batea* n. g., *Amphilochus, Atylus, Microdeutopus, Leucothoë, Melita,
Gammarus* (according to Meissner and La Valette), *Amphithoë, Cerapus,
Cyrtophium, Corophium, Dulichia, Protella* and *Caprella.*

[11] Little as a name may actually affect the facts, we ought certainly
to confine the name " micropyle" to canals of the egg-membrane, which
serve for the entrance of the semen.   But the outer egg-membrane
passes over the " micropylar apparatus" of the Amphipoda without
any perforation, according to Meissner's and La Valette's own statements; it appears never to be present before fecundation, attains its
greatest development at a subsequent period of the ovular life, and the
delicate canals which penetrate it do not even seem to be always present, indeed it seems to belong to the embryo rather than to the egg-
membrane.   I have never been able to convince myself that the so-
called " inner egg-membrane " is really of this nature, and not perhaps
the earliest larva skin, not formed until after impregnation, as might
be supposed with reference to *Ligia, Cassidina* and *Philoscia.*

phium dentatum, n. sp., and the last abdominal segments
and the tail in *Brachyscelus*,[12] or where one or more
segments are deficient, as in *Dulichia* and the *Caprellæ*,
we find the same fusion and the same deficiencies in
young animals taken out of the brood-pouch of their
mother. Even peculiarities in the structure of the
limbs, so far as they are common to both sexes, are
usually well-marked in the newly hatched young, so
that the latter generally differ from their parents only
by their stouter form, the smaller number of the antennal joints and olfactory filaments, and also of the
setæ and teeth with which the body or feet are armed,
and perhaps by the comparatively larger size of the
secondary flagellum. An exception to this rule is presented by the Hyperinæ which usually live upon Acalephæ. In these the young and adults often have a
remarkably different appearance; but even in these
there is no new formation of body-segments and limbs,
but only a gradual transformation of these parts.[13]

[12] According to Spence Bate, in *Brachyscelus crusculum* the fifth
abdominal segment is not amalgamated with the sixth the tail) but
with the fourth, which I should be inclined to doubt, considering the
close agreement which this species otherwise shows with the two
species that I have investigated.

[13] In the young of *Hyperia galba* Spence Bate did not find any of the
abdominal feet, or the last two pairs of thoracic feet, but this very
remarkable statement required confirmation the more because he
examined these minute animals only in the dried state. Subsequently
I had the wished-for opportunity of tracing the development of a
Hyperia which is not uncommon upon Ctenophora, especially *Beroë
gilva*, Eschsch. The youngest larvæ from the brood-pouch of the
mother already possess *the whole* of the thoracic feet; on the other
hand, like Spence Bate, I cannot find those of the abdomen. At first
simple enough, all these feet soon become converted, like the anterior

Thus, in order to give a few examples, the powerful chelæ of the antepenultimate pair of feet, of *Phromina*

feet, into richly denticulated prehensile feet, and indeed of three different forms, the anterior feet (fig. 44) the two following pairs (fig. 45) and finally the three last pairs (fig. 46) being similarly constructed and different from the rest. In this form the feet remain for a very long time, whilst the abdominal appendages grow into powerful natatory organs, and the eyes, which at first seemed to me to be wanting, into large hemispheres. In the transition to the form of the adult animal the last three pairs of feet (fig. 49) especially undergo a con-

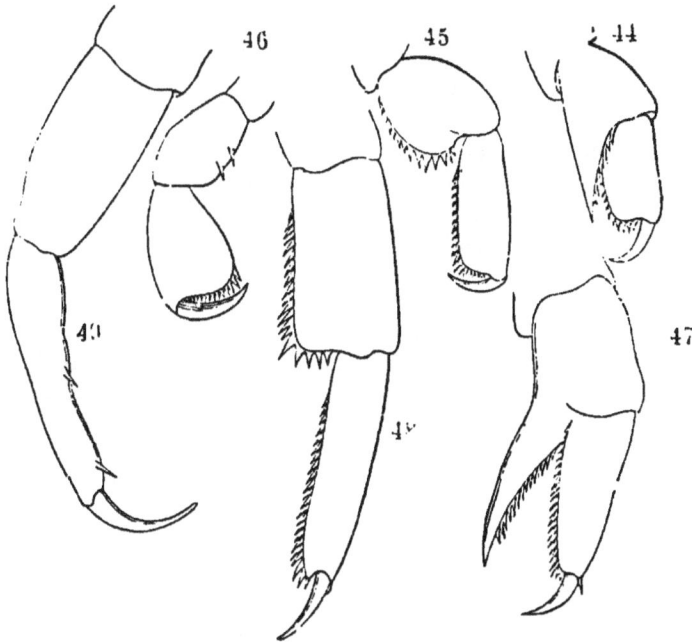

Figs. 44—49.ª

ª Figs. 44—46. Feet of a half-grown *Hyperia Martinezii*, n. sp.ᵇ Figs. 47—49. Feet of a nearly adult male of the same species; 44 and 47 from the first pair of anterior feet (gnathopoda); 44 and 48 from the first, and 46 and 49 from the last pair of thoracic feet. Magn. 90 diam.

ᵇ Named after my valued friend the amiable Spanish zoologist, M. Francisco de Paula Martinez y Saes, at present on a voyage round the world.

sedentaria, are produced, according to Pagenstecher, from simple feet of ordinary structure ; and *vice versâ*, the chelæ on the penultimate pair of feet of the young *Brachyscelus*, become converted into simple feet. In the young of the last-mentioned genus the long head is drawn out into a conical point and bears remarkably small eyes ; in course of growth, the latter, as in most of the Hyperinæ, attain an enormous size, and almost entirely occupy the head, which then appears sphe-rical, &c.

The difference of the sexes which, in the Gammarinæ is usually expressed chiefly in the structure of the

siderable change. The difference between the two sexes is considerable; the females are distinguished by a very broad thorax, and the males (*Lestrigonus*) by very long antennæ, of which the anterior bear an unusual abundance of olfactory filaments.

Their youngest larvæ of course cannot swim ; they are helpless little animals which firmly cling especially to the swimming laminæ of their host ; the adult *Hyperiæ*, which are not unfrequently met with free in the sea, are, as is well known, the most admirable swimmers in their order. ("Il nage avec une rapidité extrême," says Van Beneden of *H. Latreillii* M.-Edw.)

The transformation of the *Hyperiæ* is evidently to be regarded as *acquired* and not *inherited*, that is to say the late appearance of the abdominal appendages and the peculiar structure of the feet in the young are not to be brought into unison with the historical development of the Amphipoda, but to be placed to the account of the parasitic mode of life of the young.

As in *Brachyscelus*, free locomotion has been continued to the adult and not to the young, contrary to the usual method among parasites. Still more remarkable is a similar circumstance in *Caligus*, among the parasitic Copepoda. The young animal, described by Burmeister as a peculiar genus, *Chalimus*, lies at anchor upon a fish by means of a cable springing from its forehead, and having its extremity firmly seated in the skin of the fish. When sexual maturity is attained, the cable is cut, and the adult *Caligi*, which are admirable swimmers, are not unfrequently captured swimming freely in the sea. (See ' Archiv. für Naturg.' 1852, I. p. 91).

anterior feet (gnathopoda, Sp. Bate) and in the Hyperinæ in the structure of the antennæ, is often so great that males and females have been described as distinct species, and even repeatedly placed in different genera (*Orchestia* and *Talitrus*, *Cerapus* and *Dercothoë*, *Lestrigonus* and *Hyperia*) or even families (*Hypérines anormales* and *Hypérines ordinaires*). Nevertheless it is only developed when the animals are nearly full-grown. Up to this period the young resemble the females in a general way, even in some cases in which these differ more widely than the males from the "Type" of the order. Thus in the male Shore-hoppers (*Orchestia*) the second pair of the anterior feet is provided with a powerful hand, as in the majority of the Amphipoda, but very differently constructed in the females. The young, nevertheless, resemble the female. Thus also,— and this is an extremely rare case,[14]—the females of *Brachyscelus* are destitute of the posterior (or inferior) antennæ; the male possesses them like other Amphipodæ; in the young I, like Spence Bate, can find no trace of them.

It is, however, to be particularly remarked, that the development of the sexual peculiarities does not stand still on the attainment of sexual maturity.

For example, the younger sexually mature males of *Orchestia Tucurauna*, n. sp., have slender inferior antennæ, with the joints of the flagellum not fused together, the clasping margin (" palm," Sp. Bate) of the

[14] "I know of no case in which the inferior (antennæ) are obsolete, when the superior are developed," Dana. (Darwin, ' Monograph on the Subclass Cirripedia, Lepadidæ,' p. 15.)

hand in the second pair of feet is uniformly convex, the last pair of feet is slender and similar
to the preceding. Subsequently the antennæ become
thickened, two, three, or four of the first joints of the
flagellum are fused together, the palm of the hand
acquires a deep emargination near its inferior angle,
and the intermediate joints of the last pair of feet
become swelled into a considerable incrassation. No
museum-zoologist would hesitate about fabricating two
distinct species, if the oldest and youngest sexually
mature males were sent to him without the uniting intermediate forms. In the younger males of *Orchestia
Tucuratinga*, although the microscopic examination of

Fig. 50.[15]

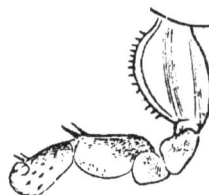

Fig. 51.[15]

their testes showed that they were already sexually
mature, the emargination of the clasping margin of the
hand (represented in fig. 50) and the corresponding process of the finger, are still entirely wanting. The same
may be observed in *Cerapus* and *Caprella*, and probably
in all cases where hereditary sexual differences occur.

[15] Fig. 50. Foot of the second pair (" second pair of gnathopoda ") of
the male and fig. 51 of the female, of *Orchestia Tucuratinga*, magn. 15
diam.

Next to the extensive sections of the Stalk-eyed and Sessile-eyed Crustacea, but more nearly allied to the former than to the latter, comes the remarkable family of the *Diastylidæ* or *Cumacea.* The young, which Kröyer took out of the brood-pouch of the female, and which attained one-fourth of the length of

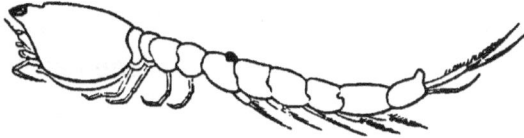

Fig. 52. [16]

their mother, resembled the adult animals almost in all parts. Whether, as in *Mysis* and *Ligia*, a transformation occurs within the brood-pouch, which is constructed in the same way as in *Mysis*, is not known.[17]   The caudal

[16] Fig. 52. Male of a *Bodotria*, magn. 10 diam. Note the long inferior antenuæ, which are closely applied to the body, and of which the apex is visible beneath the caudal appendages.

[17] A trustworthy English Naturalist, Goodsir, described the brood-pouch and eggs of *Cuma* as early as 1843. Kröyer, whose painstaking care and conscientiousness is recognised with wonder by every one who has met him on a common field of work, confirmed Goodsir's statements in 1846, and, as above mentioned, took out of the brood-pouch embryos advanced in development and resembling their parents. By this the question whether the Diastylidæ are full-grown animals or larvæ, is completely and for ever set at rest, and only the famous names of Agassiz, Dana and Milne-Edwards, who would recently reduce them again to larvæ (see Van Beneden, 'Rech. sur la Faune littor. de Belgique,' Crustacées, pp. 73, 74), induce me, on the basis of numerous investigations of my own, to declare in Van Beneden's words ; "Parmi toutes les formes embryonnaires de podophthalmes ou d'édriophthalmes que nous avons observées sur nos côtes, nous n'en avons pas vu une seule qui eût même la moindre ressemblance avec un *Cuma* quelconque." The *only thing* that suits the larvæ of *Hippolyte, Palæmon* and *Alpheus*, in the family character of the Cumacea as given by Kröyer which occupies three pages (Kröyer, 'Naturh. Tidsskrift, Ny Række,' Bd. ii. pp. 203—206, is : "Duo antennarum paria." And this, as is well known

portion of the embryo in the *Diastylidæ*, as I have recently observed, is curved upwards as in the Isopoda, and the last pair of feet of the thorax is wanting.

Equally scanty is our knowledge of the developmental history of the Ostracoda. We know scarcely anything except that their anterior limbs are developed before the posterior one (Zenker). The development of *Cypris* has recently been observed by Claus :—" The youngest stages are shell-bearing Nauplius-forms."

applies to nearly all Crustacea. How well warranted are we therefore in identifying the latter with the former. However, it is sufficient for any one to glance at the larva of *Palæmon* (fig. 27) and the Cumacean (fig. 52) in order to be convinced of their extraordinary similarity !

CHAPTER IX.

DEVELOPMENTAL HISTORY OF ENTOMOSTRACA, CIRRI-PEDES, AND RHIZOCEPHALA.

THE section of the Branchiopoda includes two groups differing even in their development,—the Phyllopoda and the Cladocera. The latter minute animals, provided with six pairs of foliaceous feet, which chiefly belong to the fresh waters, and are diffused under similar forms over the whole world, quit the egg with their full number of limbs. The Phyllopoda, on the contrary, in which the number of feet varies between 10 and 60 pairs, and some of which certainly live in the saturated lie of salterns and natron-lakes, but of which only one rather divergent genus (*Nebalia*) is found in the sea,[1] have to undergo a metamorphosis. Mecznikow has recently observed the development of *Nebalia*, and concludes from his observations "that *Nebalia*, during its embryonal life, passes through the

---

[1] If the Phyllopoda may be regarded as the nearest allies of the Trilobites, they would furnish, with *Lepidosteus* and *Polypterus*, *Lepidosiren* and *Protopterus*, a further example of the preservation in fresh waters of forms long since extinguished in the sea. The occurrence of the *Artemiæ* in supersaline water would at the same time show that they do not escape destruction by means of the fresh water, but in consequence of the less amount of competition in it.

Nauplius- and Zoëa-stages, which in the Decapoda occur partly (in *Penëus*) in the free state." "Therefore," says he, "I regard *Nebalia* as a Phyllopodiform Decapod." The youngest larvæ [of the Phyllopoda] are Nauplii, which we have already met with exceptionally in some Prawns, and which we shall now find reproduced almost without exception. The body-segments and feet, which are sometimes so numerous, are formed gradually from before backwards, without the indication of any sharply-discriminated regions of the body either by the time of their appearance or by their form. All the feet are essentially constructed in the same manner and resemble the maxillæ of the higher Crustacea.[2] We might regard the Phyllopoda as Zoëæ which have not arrived at the formation of a peculiarly endowed abdomen or thorax, and instead of these have repeatedly reproduced the appendages which first follow the Nauplius-limbs.

Of the Copepoda—some of which, living in a free state, people the fresh waters, and in far more multifarious forms the sea, whilst others, as parasites, infest animals of the most various classes and often become wonderfully deformed—the developmental history, like their entire natural history, was, until lately, in a very unsatisfactory state. It is true, that we long ago knew that the *Cyclopes* of our fresh waters were excluded in the Nauplius-form, and that we were acquainted with some others of their young states; we had learnt,

[2] "The maxilla of the Decapod-larva (Krebslarve) is a sort of Phyllopodal foot" (Claus).

through Nordmann, that the same earliest form be-
longed to several parasitic Crustacea, which had previously passed, almost universally, as worms; but the
connecting intermediate forms which would have per-
mitted us to refer the regions of the body and the limbs
of the larvæ to those of the adult animal, were wanting.
The comprehensive and careful investigations of Claus
have filled up this deficiency in our knowledge, and
rendered the section of the Copepoda one of the best
known in the whole class. The following statements
are derived from the works of this able naturalist.
From the abundance of valuable materials which they
contain I select only those which are indispensable for
the comprehension of the development of the Crustacea
in general, because, in what relates to the Copepoda in
particular, the facts have already been placed in the
proper light by the representation of their most recent
investigator, and must appear to any one whose eyes
are open, as important evidence in favour of the Darwinian theory.[3]

All the larvæ of the free Copepoda investigated by
Claus, have, at the earliest period, three pairs of limbs
(the future antennæ and mandibles), the anterior with a
single, and the two following ones with a double series
of joints, or branchiæ. The unpaired eye, labrum, and
mouth, already occupy their permanent positions. The
posterior portion, which is usually short and destitute of
limbs, bears two terminal setæ, between which the anus

---

[3] I am still unacquainted with Claus' latest and larger work, but no
doubt the same may be said of it.

is situated. The form in this Nauplius-brood is extremely various,—it is sometimes compressed laterally, sometimes flat,—sometimes elongated, sometimes oval, sometimes round or even broader than long, and so forth. The changes which the first larval stages undergo during the progress of growth, consist essentially in an extension of the body and the sprouting forth of new limbs. "The following stage already displays a

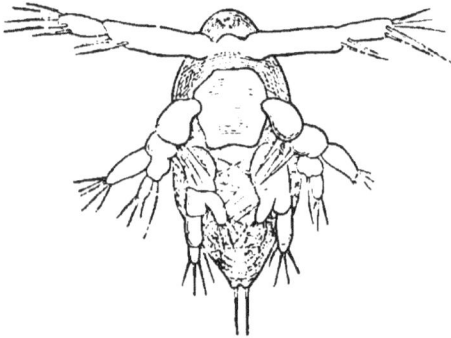

Fig. 53.[4]

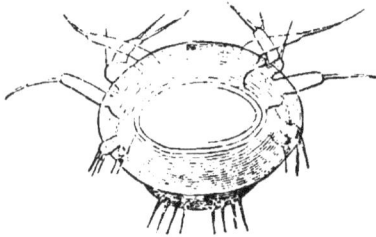

Fig. 54.[4]

fourth pair of extremities, the future maxillæ." Then follow at once three new pairs of limbs (the maxillipedes and the two anterior pairs of natatory feet). The larva still continues like a Nauplius, as the three anterior pairs of limbs represent rowing feet ; at the next moult it is converted into the youngest *Cyclops*-like state, when it resembles the adult animal in the structure of the antennæ and buccal organs, although the number of limbs and body-segments is still much less, for only the rudiments of the third and fourth pairs of natatory feet have made their appearance in the form of cushions fringed with setæ, and the body consists of the oval cephalothorax, the second, third, and fourth thoracic segments, and an elongated terminal joint. In the Cyclopidæ the posterior antennæ have lost their secondary branch, and the mandibles have completely thrown off the previously existing natatory feet, whilst in the other families these appendages persist, more or less altered. "Beyond this stage of free development, many forms of the parasitic Copepoda, such as *Lernanthropus* and *Chondracanthus*, do not pass, as they do not acquire the third and fourth pairs of limbs, nor does a separation of the fifth thoracic segment from the abdomen take place; others (*Achtheres*) even fall to a lower grade by the subsequent loss of the two pairs of natatory feet. But all free Copepoda, and most of the parasitic Crustacea, pass through a longer or shorter series of stages of development, in which the limbs acquire a higher degree of division into joints in continuous sequence, the posterior pairs of feet are developed, and the last thoracic segment and the different abdominal segments are successively separated from the common terminal portion."

There is only one thing more to be indicated in the developmental history of the parasitic Crustacea, namely that some of them, such as *Achtheres percarum*, certainly quit the egg like the rest in a Nauplius-like form, inasmuch as the plump, oval, astomatous body bears two pairs of simple rowing feet, and behind these, as traces

of the third pair, two inflations furnished each with a
long seta, but that beneath this Nauplius-skin a very
different larva lies ready prepared, which in a few hours
bursts its clumsy envelope and then makes its appearance
in a form " which
agrees in the seg-
mentation of the
body and in the de-
velopment of the ex-
tremities with the
first *Cyclops-stage*"
(Claus). The en-
tire series of Nau-
plius-stages which
are passed through
by the free Copepo-
da, are in this case
completely over-
leapt.

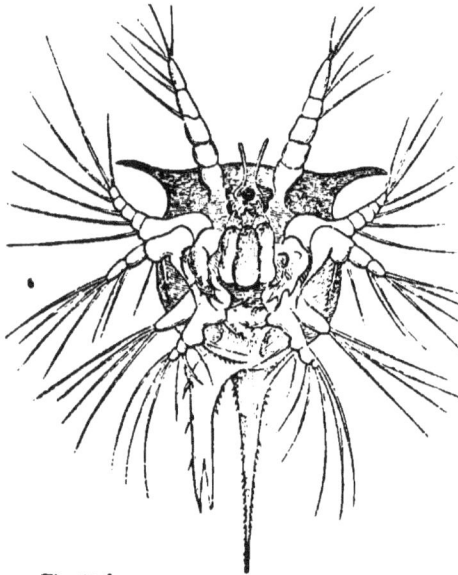

Fig. 55. ⁵

A final and very peculiar section of the Crustacea is
formed by the two orders of the Cirripedia and Rhizo-
cephala.⁶

⁵ Fig. 55. Nauplius of *Tetraclita porosa* after the first moult, magn.
90 diam. The brain is seen surrounding the eye, and from it the
olfactory filaments issue ; behind it are some delicate muscles passing
to the buccal hood.

⁶ The most various opinions prevail as to the position of the Cirripedia.
Some ascribe to them a very subordinate position among the Copepoda;
as Milne-Edwards (1852). In direct opposition to this notion of his
father's, Alph. Milne-Edwards places them (as *Basinotes*) opposite to
all the other Crustacea (*Eleuthéronotes*). Darwin regards them as
forming a peculiar sub-class equivalent to the Podophthalma, Edriophthalma, &c. This appears to me to be most convenient. I would not

In these also the brood bursts out in the Nauplius-form, and speedily strips off its earliest larva-skin which is distinguished by no peculiarities worth noticing. Here also we find again the same pyriform shape of the un-segmented body, the same number and structure of the feet, the same position of the median eye (which, how-ever, is wanting in *Sacculina purpurea*, and according to Darwin in some species of *Lepas*), and the same position of the "buccal hood," as in the Nauplii of the Prawns and Copepoda. From the latter the Nauplii of the Cirripedia and Rhizocephala are distinguished by the possession of a dorsal shield or carapace, which sometimes (*Sacculina purpurea*) projects far beyond the body all round; and they are distinguished not only from other Nauplii, but as far as I know from all other Crustacea, by the circumstance that structures which are elsewhere combined with the two anterior limbs (antennæ), here occur separated from them.

The anterior antennæ of the Copepoda, Cladocera, Phyllopoda (Leydig, Claus), Ostracoda (at least the Cypridinæ), Diastylidæ, Edriophthalma, and Podoph-thalma, with few exceptions relating to terrestrial ani-mals or parasites, bear peculiar filaments which I have already repeatedly mentioned as " olfactory filaments."

combine the Rhizocephala with the Cirripedia, as Liljeborg has done, but place them in opposition as equivalent, like the Amphipoda and Isopoda. The near relationship of the Cirripedia to the Ostracoda is also spoken of, but the similarity of the so-called " *Cypris*-like larvæ," or Cirriped-pupæ as Darwin calls them, to *Cypris* is so purely external, even as regards the shell, that the relationship appears to me to be scarcely greater than that of *Peltogaster socialis* (fig. 59) with the family of the sausages.

A pair of similar filaments spring, in the larvæ of the Cirripedia and Rhizocephala, directly from the brain.

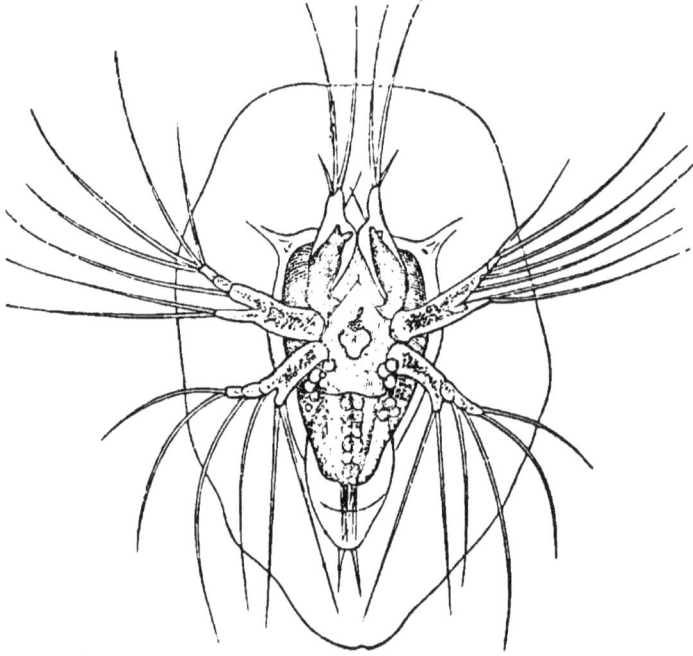

Fig. 56.[7]

At the base of the inferior antennæ in the Decapoda the so-called "green-gland" has its opening; in the Macrura at the end of a conical process. A similar conical process with an efferent duct traversing it is very striking in most of the Amphipoda. In the Ostracoda, Zenker describes a gland situated in the base of the inferior antennæ, and opening at the extremity of an extraordinarily long "spine." In the Nauplii of *Cyclops*

---

[7] Fig. 56. Nauplius of *Sacculina purpurea*, shortly before the second moult, magn. 180 diam. We may recognise in the first pair of feet the future adherent feet, and in the abdomen six pairs of natatory feet with long setæ.

and *Cyclopsine*, Claus finds pale "shell-glands," which commence in the intermediate pair of limbs (the posterior antennæ). On the other hand in the Nauplii of the Cirripedia and Rhizocephala the "shell-glands" open at the ends of conical processes, sometimes of most remarkable length, which spring from the angles of the broad frontal margin, and have been interpreted sometimes as antennæ (Burmeister, Darwin) and sometimes as mere "horns of the carapace" (Krohn). The connexion of the "shell-glands" with the frontal horns has been recognised unmistakably in the larvæ of *Lepas*, and indeed the resemblance of the frontal horns with the conical processes on the inferior antennæ of the Amphipoda, is complete throughout.[8]

Notwithstanding their agreement in this important peculiarity, the Nauplii of these two orders present material differences in many other particulars. The abdomen of the young Cirripede is produced beneath the anus into a long tail-like appendage which is furcate at the extremity, and over the anus there is a second long, spine-like process; the abdomen in the Rhizocephala terminates in two short points,—in a "moveable caudal fork, as in the Rotatoria," (O. Schmidt). The young Cirripedes have a mouth, stomach, intestine, and anus, and their two posterior pairs of limbs are beset with multifarious teeth, setæ, and hooks, which certainly assist in the inception of nourishment. All this is wanting in the young Rhizo-

---

[8] In connexion with this it may be mentioned that, in the females of *Brachyscelus*, in which the posterior antennæ are deficient, the conical processes with the canal permeating them are nevertheless retained.

cephala. The Nauplii of the Cirripedia have to undergo several moults whilst in that form; the Nauplii of the Rhizocephala, being astomatous, cannot of course live long as Nauplii, and in the course of only a few days they become transformed into equally astomatous "pupæ," as Darwin calls them.

The carapace folds itself together, so that the little animal acquires the aspect of a bivalve shell, the foremost limbs become transformed into very peculiar adherent feet ("prehensile antennæ," Darwin), and the two following pairs are cast off, like the frontal horns. On the abdomen six pairs of powerful biramose natatory feet with long setæ have been formed beneath the Nauplius-skin, and behind these are two short, setigerous caudal appendages (fig. 58).

Fig. 57.⁹

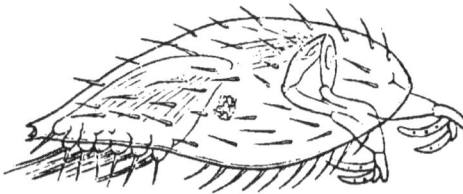

Fig. 58. ¹⁰

The pupæ of the Cirripedia (fig. 57), which are likewise astomatous, agree completely in all these parts with those of the Rhizocephala, even to the minutest details of the segmenta-

⁹ Fig. 57. Pupa of a Balanide (*Chthamalus?*), magn. 50 diam. The adherent feet are retracted within the rather opaque anterior part of the shell.

¹⁰ Fig. 58. Pupa of *Sacculina purpurea*, magn. 180 diam. The filaments on the adherent feet may be the commencements of the future roots.

tion and bristling of the natatory feet;[11] they are especially distinguished from them by the possession of a pair of composite eyes. Sometimes also traces of the frontal horns seem to persist.[12]

As the Cirripedia and Rhizocephala now in general resemble each other far more than in their Nauplius-state, this is also the case with the individual members of each of the two orders.

The pupæ in both orders attach themselves by means of the adherent feet; those of the Cirripedes to rocks, shells, turtles, drift-wood, ships, &c.,—those of the Rhizocephala to the abdomen of Crabs, *Porcellanæ*, and Hermit Crabs. The carapace of the Cirripedes becomes converted, as is well-known, into a peculiar test, on account of which they were formerly placed among the Mollusca, and the natatory feet grow into long cirri, which whirl nourishment towards the mouth, which is now open. The Rhizocephala remain astomatous; they lose **all** their limbs completely, and appear as sausage-like, sack-shaped or discoidal excrescences of their host, filled with ova (figs. 59, 60); from the point of

[11] Compare the figure given by Darwin (Balanidæ, Pl. xxx. fig. 5) of the first natatory foot of the pupa of *Lepas australis*, with that of *Lernæodiscus Porcellanæ* published in the 'Archiv für Naturgeschichte' (1863, Taf. iii. fig. 5). The sole distinction, that in the latter there are only 3 setæ at the end of the outer branch, whilst in the Cirripedia there are 4 on the first and 5 on the following natatory feet, may be due to an error on my part.

[12] Darwin describes as "acoustic orifices" small apertures in the shell of the pupæ of the Cirripedia, which, frequently surrounded by a border, are situated, in *Lepas pectinata*, upon short, hornlike processes. I feel scarcely any hesitation in regarding the apertures as those of the "shell-glands," and the hornlike processes as remains of the frontal horns.

attachment closed tubes, ramified like roots, sink into the interior of the host, twisting round its intestine, or becoming diffused among the sac-like tubes of its liver. The only manifestations of life which

persist in these *non plus ultras* in the series of retrogressively metamorphosed Crustacea, are powerful contractions of the roots, and an alternate expansion and contraction of the

Fig. 59.[13]   Fig. 60.[14]

body, in consequence of which water flows into the brood-cavity and is again expelled, through a wide orifice.[15]

Out of several Cirripedes, which are anomalous both in structure and development, *Cryptophialus minutus* must be mentioned here ; Darwin found it in great quantities together in the shell of *Concholepas peruviana* on the

[13] Fig. 59. Young of *Peltogaster socialis* on the abdomen of a small Hermit Crab ; in one of them the fasciculately ramified roots in the liver of the crab are shown. Animal and roots deep yellow.

[14] Fig. 60. Young *Sacculina purpurea* with its roots ; the animal purple-red, the roots dark grass-green. Magn. 5 diam.

[15] The roots of *Sacculina purpurea* (fig. 60) which is parasitic upon a small Hermit Crab, are made use of by two parasitic Isopods, namely a *Bopyrus* and the before mentioned *Cryptoniscus planarioïdes* (fig. 42). These take up their abode beneath the *Sacculina* and cause it to die away by intercepting the nourishment conveyed by the roots; the roots, however, continue to grow, even without the *Sacculina*, and frequently attain an extraordinary extension, especially when a *Bopyrus* obtains its nourishment from them.

Chonos Islands. The egg, which is at first elliptical, soon, according to Darwin, becomes broader at the anterior extremity, and acquires three club-shaped horns, one at each anterior angle and one behind; no internal parts can as yet be detected. Subsequently the posterior horn disappears, and the adherent feet may be recognised within the anterior ones. From this "egg-like larva"—(Darwin says of it, "I hardly know what to call it")—the pupa is directly produced. Its carapace is but slightly compressed laterally and hairy, as in *Sacculina purpurea;* the adherent feet are of considerable size, and the natatory feet are wanting, as, in the adult animal, are the corresponding cirri. As I learn from Mr. Spence Bate, the Nauplius-stage appears to be overleaped and the larvæ to leave the egg in the pupa-form, in the case of a Rhizocephalon (*Peltogaster ?*) found by Dr. Powell in the Mauritius.

I will conclude this general view with a few words upon the earliest processes in the development of the Crustacea. Until recently it was regarded as a general rule that, by the partial segmentation of the vitellus a germinal disc was formed, and in this, cor-

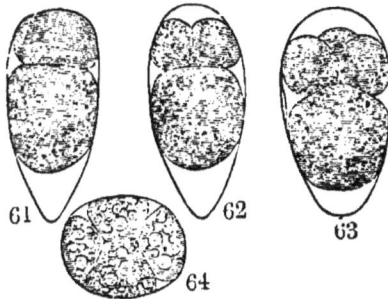

Figs. 61, 62, 63, 64. [16]

[16] Figs. 61—63. Eggs of *Tetraclita porosa* in segmentation, magn. 90 diam.   The larger of the two first-formed spheres of segmentation is always turned towards the pointed end of the egg. Fig. 64. Egg of *Lernæodiscus Porcellanæ*, in segmentation, magn. 90 diam.

responding to the ventral surface of the embryo, a primitive band. We now know that in the Copepoda (Claus),
in the Rhizocephala (fig. 64), and, as I can add, in the
Cirripedia (figs. 61-63) the segmentation is complete, and
the embryos are sketched out in their complete form
without any preceding primitive band. Probably the
latter will always be the case where the young are
hatched as true *Nauplii* (and not merely with a Nauplius-skin, as in *Achtheres*). The two modes of development may occur in very closely allied animals, as
is proved by *Achtheres* among the Copepoda.[17]

[17] I have not mentioned the Pycnogonidæ, because I do not regard
them as Crustacea ; nor the Xiphosura and Trilobites, because, having
never investigated them myself, I knew too little about them, and
especially because I am unacquainted with the details of the explanations given by Barrande of the development of the latter. According
to Mr. Spence Bate " the young of Trilobites are of the Nauplius-
form."

CHAPTER X.

ON THE PRINCIPLES OF CLASSIFICATION.

PERHAPS some one else, more fortunate than myself, may be able, even without Darwin, to find the guiding clue through the confusion of developmental forms, now so totally different in the nearest allies, now so surprisingly similar in members of the most distant groups, which we have just cursorily reviewed. Perhaps a sharper eye may be able, with Agassiz, to make out " the plan established from the beginning by the Creator," [1] who may have written here, as a Portuguese proverb says " straight in crooked lines." [2] I cannot but think that we can scarcely speak of a general plan, or typical mode of development of the Crustacea, differentiated according to the separate Sections, Orders, and Families, when, for example, among the Macrura, the River Crayfish leaves the egg in its permanent form ; the

[1] "A plan fully matured in the beginning and undeviatingly pursued ;" or " In the beginning His plan was formed and from it He has never swerved in any particular" (Agassiz and Gould, 'Principles of Zoology').

[2] " Deos escrive direito em linhas tortas." To read this remarkable writing we need the spectacles of Faith, which seldom suit eyes accustomed to the Microscope.

Lobster with Schizopodal feet ; *Palæmon*, like the Crabs, as a Zoëa ; and *Penëus*, like the Cirripedes, as a Nauplius,—and when, still, within this same sub-order Macrura, *Palinurus, Mysis* and *Euphausia* again present different young forms,—when new limbs sometimes sprout forth as free rudiments on the ventral surface, and are sometimes formed beneath the skin which passes smoothly over them, and both modes of development are found in different limbs of the same animal and in the same pair of limbs in different animals,—when in the Podophthalma the limbs of the thorax and abdomen make their appearance sometimes simultaneously, or sometimes the former and sometimes the latter first, and when further in each of the two groups the pairs sometimes all appear together, and sometimes one after the other,—when, among the Hyperina, a simple foot becomes a chela in *Phronima* and a chela a simple foot in *Brachyscelus*, &c.

And yet, according to the teaching of the school, it is precisely in youth, precisely in the course of development, that the "Type" is mostly openly displayed. But let us hear what the Old School has to tell us as to the significance of developmental history, and its relation to comparative anatomy and systematic zoology.

Let two of its most approved masters speak.

"Whilst comparative anatomy," said Johannes Müller, in 1844, in his lectures upon this science (and the opinions of my memorable teacher were for many years my own), "whilst comparative anatomy shows us the infinitely multifarious formation of the same organ

in the Animal Kingdom, it furnishes us at the same
time with the means, by the comparison of these various
forms, of recognising the truly essential, the type of
these organs, and separating therefrom everything un-
essential. In this, developmental history serves it as a
check or test. Thus, as the idea of development is not
that of mere increase of size, but that of progress from
what is not yet distinguished, but which potentially
contains the distinction in itself, to the actually distinct,—it is clear, that the less an organ is developed, so
much the more does it approach the type, and that,
during its development, it more and more acquires
peculiarities. The types discovered by comparative
anatomy and developmental history must therefore
agree."

Then, after Johannes Müller has combated the idea
of a graduated scale of animals, and of the passage
through several animal grades during development, he
continues:—" What is true in this idea is, that every
embryo at first bears only the type of its section, from
which the type of the Class, Order, &c., is only afterwards developed."

In 1856, in an elementary work,[3] in which it is usual
to admit only what are regarded as the assured acquisi-
tions of science, Agassiz expresses himself as follows :—

" *The ovarian eggs of all animals are perfectly identical*, small cells with a vitellus, germinal vesicle and
germinal spot" (§ 278). " *The organs of the body are*

[3] ' Principles of Zoology.' Part I. Comparative Physiology. By Louis
Agassiz and A. A. Gould. Revised Edition. Boston, 1856.

*formed in the sequence of their organic importance; the
most essential always appear first.* Thus the organs of
vegetative life, the intestine, &c., appear later than
those of animal life, the nervous system, skeleton, &c.,
and these in turn are preceded by the more general
phenomena belonging to the animal as such " (§ 318).
" Thus, in Fishes, the first changes consist in the segmentation of the vitellus and the formation of a germ,
processes which are common to all classes of animals.
Then the dorsal furrow, characteristic of the Vertebrate,
appears—the brain, the organs of the senses; at a later
period are formed the intestine, the limbs, and the permanent form of the respiratory organs, from which the
class is recognised with certainty. It is only after exclusion that the peculiarities of the structure of the
teeth and fins indicate the genus and species " (§ 319).
" *Hence the embryos of different animals resemble each
other the more, the younger they are* " (§ 320). " Consequently the high importance of developmental history
is indubitable. For, *if the formation of the organs takes
place in the order corresponding to their importance, this
sequence must of itself be a criterion of their comparative value in classification.* The peculiarities which
appear earlier should be considered of higher value
than those which appear subsequently " (§ 321). " *A
system, in order to be true and natural, must agree with
the sequence of the organs in the development of the
embryo* " (§ 322).

I do not know whether any one at the present day
will be inclined to subscribe to this proposition in its

whole extent.[4] It is certain, however, that views essentially similar are still to be met with everywhere in
discussions on classification, and that even within the
last few years, the very sparingly successful attempts
to employ developmental history as the foundation
of classification have been repeated.

But how do these propositions agree with our observations on the developmental history of the Crustacea?
That these observations relate for the most part to
their "free metamorphosis" after their quitting the
egg, cannot prejudice their application to the propositions enunciated especially with regard to "embryonal
development" in the egg; for Agassiz himself points
out (§ 391) that both kinds of change are of the same
nature and of equal importance and that no "radical
distinction" is produced by the circumstance that the
former take place before and the latter after birth.

"*The ovarian eggs of all animals are identical*, small
cells with vitellus, germinal vesicle and germinal
spot." Yes, somewhat as all Insects are identical,
small animals with head, thorax, and abdomen; that is
to say if, only noticing what is common to them, we
leave out of consideration the difference of their development, the presence or absence and the multifa-

---

[4] Agassiz' own views have lately become essentially different, so far
as can be made out from Rud. Wagner's notice of his 'Essay on Classification.' Agassiz himself does not attempt any criticism of the above
cited older views, which, however, are still widely diffused. With his
recent conception I am unfortunately acquainted only from R. Wagner's
somewhat confused report, and have therefore thought it better not to
attempt any critical remarks upon it.

rious structure of the vitelline membrane, the varying composition of the vitellus, the different number and formation of the germinal spots, &c. Numerous examples, which might easily be augmented, of such profound differences, are furnished by Leydig's ' Lehrbuch der Histologie.' In the Crustacea the ovarian eggs actually sometimes furnish excellent characters for the discrimination of species of the same genus ; thus, for example, in one *Porcellana* of this country they are blackish-green, in a second deep blood-red, and in a third dark yellow ; and within the limits of the same order they present considerable differences in size, which, as Van Beneden and Claus have already pointed out, stands in intimate connexion with the subsequent mode of development.

" *The organs of the body are formed in the sequence of their organic importance ; the most essential always appear first.*" This proposition might be characterised *à priori* as undemonstrable, since it is impossible either in general, or for any particular animal, to establish a sequence of importance amongst equally indispensable parts. Which is the more important, the lung or the heart ?—the liver or the kidney ?—the artery or the vein ? Instead of giving the preference, with Agassiz, to the organs of animal life, we might with equal justice give it to those of vegetative life, as the latter are conceivable without the former, but not the former without the latter. We might urge that, according to this proposition, provisional organs as the first produced must exceed the later-formed permanent organs in importance.

But let us stick to the Crustacea. In *Polyphemus*
Leydig finds the first traces of the intestinal tube even
during segmentation. In *Mysis* a provisional tail is
first formed, and in *Ligia* a maggot-like larva-skin.
The simple median eye appears earlier, and would
therefore be more important than the compound paired
eyes; the scale of the antennæ in the Prawns would
be more important than the flagellum; the maxillipedes of the Decapoda would be more important than
the chelæ and ambulatory feet, and the anterior six
pairs of feet in the Isopoda, than the precisely similarly
formed seventh pair; in the Amphipoda the most important of all organs would be the "micropylar apparatus," which disappears without leaving a trace soon
after hatching; in *Cyclops* the setæ of the tail would
be more important than all the natatory feet; in the
Cirripedia the posterior antennæ, as to which we do not
know what becomes of them, would be more important
than the cirri, and so forth. The most unimportant of
all organs would be the sexual organs, and the most
essential peculiarity would consist in colour, which is to
be referred back to the ovarian egg.

"*The embryos, or young states of different animals,
resemble each other the more, the younger they are,*" or, as
Johannes Müller expresses it, "*they approach the more
closely to the common type.*" Different as may be the
ideas connected with the word "type," no one will dis-
pute that the typical form of the penultimate pair of
feet in the Amphipoda is that of a simple ambulatory
foot, and not that of a chela, for the latter occurs in no

single adult Amphipod; we know it only in the young
of the genus *Brachyscelus*, which therefore in this
respect undoubtedly depart more widely than the
adults from the type of their order. This applies also
to the young males of the Shore-hoppers (*Orchestia*)
with regard to the second pair of anterior feet (*gnathopoda*). In like manner no one will hesitate to accept
the possession of seven pairs of feet as a "typical"
peculiarity of the Edriophthalma, which Agassiz, on
this account, names Tetradecapoda ; the young Isopoda,
which are Dodecapoda, are also in this respect further
from the "type" than the adults.

It is certainly a rule, and this Darwin's theory would
lead us to expect, that in the progress of development
those forms which are at first similar gradually depart
further from each other ; but here, as in other classes,
the exceptions, for which the Old School has no explanation, are numerous. Not unfrequently we might
indeed directly reverse the proposition and assert that
the difference becomes the greater, the further we go
back in the development, and this not only in those
cases in which one of two nearly allied species is directly developed, and the other passes through several
larval stages, such as the common Crayfish and the
Prawns which are produced from Nauplius-brood.
The same may be said, for example, of the Isopoda
and Amphipoda. In the adult animals the number of
limbs is the same ; at the first sight of a *Cyrtophium* or
a *Dulichia*, and even after the careful examination of a
Tanais, we may be in doubt whether we have an

Isopod or an Amphipod before us; in the newly-hatched young the number of limbs is different, and if we go back to their existence in the egg, the most passing glance to see whether the curvature is upwards or downwards suffices to distinguish even the youngest embryos of the two orders.

In other instances, the courses which lead from a similar starting-point to a similar goal, separate widely in the middle of the development, as in the Prawns with Nauplius-brood already described.

Finally, so that even the last possibility may be exhausted, it sometimes happens that the greatest similarity occurs in the middle of the development. The most striking example of this is furnished by the Cirripedia and Rhizocephala, whether we compare the two orders or the members of each with one another; from a segmentation quite different in its course (see figs. 61–64) proceed different forms of Nauplius, these become converted into exceedingly similar pupæ, and from the pupæ again proceed sexually mature animals, differing from each other *toto cœlo*.

" *If the formation of the organs occurs in the order corresponding to their importance, this sequence must of itself be a criterion of their comparative value in classification,*" THAT IS TO SAY, SUPPOSING THE PHYSIOLOGICAL AND CLASSIFICATIONAL VALUE OF AN ORGAN TO CO-INCIDE! Just as in Christian countries there is a catechismal morality, which every one has upon his lips, but no one considers himself bound to follow, or expects to see followed by anybody else, so also has

Zoology its dogmas, which are as universally acknow-
ledged, as they are disregarded in practice. Such a
dogma as this is the supposition tacitly made by Agassiz.
Of a hundred who feel themselves compelled to give
their systematic confession of faith as the introduction
to a Manual or Monographic Memoir, ninety-nine will
commence by saying that a natural system cannot be
founded upon a single character, but that it has to take
into account all characters, and the general structure of
the animal, but that we must not simply sum up these
characters like equivalent magnitudes, that we must not
count but weigh them, and determine the importance
to be ascribed to each of them according to its physiological significance. This is probably followed by a
little jingle of words in general terms on the comparative importance of animal and vegetative organs,
circulation, respiration, and the like. But when we
come to the work itself, to the discrimination and arrangement of the species, genera, families, &c., in all
probability not one of the ninety-nine will pay the least
attention to these fine rules, or undertake the hopeless
attempt to carry them out in detail. Agassiz, for
example, like Cuvier, and in opposition to the majority
of the German and English zoologists, regards the
Radiata as one of the great primary divisions of the
Animal Kingdom, although no one knows anything
about the significance of the radiate arrangement in
the life of these animals, and notwithstanding that the
radiate Echinodermata are produced from bilateral larvæ.
The "true Fishes" are divided by him into Ctenoids

and Cycloids, according as the posterior margin of their
scales is denticulated or smooth, a circumstance the
importance of which to the animal must be infinitely
small, in comparison to the peculiarities of the dentition,
formation of the fins, number of vertebræ, &c.

And, to return to our Class of the Crustacea, has any
particular attention been paid in their classification
to the distinctions prevailing in the "most essential
organs"? For instance, to the nervous system? In the
Corycæidæ, Claus found all the ventral ganglia fused
together into a single broad mass, and in the Calanidæ
a long ventral chain of ganglia,—the former, therefore,
in this respect resembling the Spider Crabs and the
latter the Lobster; but no one would dream on this
account of supposing that there was a relationship between the Corycæidæ and the Crabs, or the Calanidæ
and the Lobsters.—Or to the organs of circulation?
We have among the Copepoda, the Cyclopidæ and
Corycæidæ without a heart, side by side with the
Calanidæ and Pontellidæ with a heart. And in the
same way among the Ostracoda, the *Cypridinæ*, which
I find possess a heart, place themselves side by side
with *Cypris* and *Cythere* which have no such organ.—
Or to the respiratory apparatus? Milne-Edwards did
this when he separated *Mysis* and *Leucifer* from the
Decapoda, but he himself afterwards saw that this was
an error. In one *Cypridina* I find branchiæ of considerable size, which are entirely wanting in another
species, but this does not appear to me to be a reason
for separating these species even generically.

On the other hand, what do we know of the physiological significance of the number of segments, and all
the other matters which we are accustomed to regard
as typical peculiarities of the different organs, and
to which we usually ascribe the highest systematic
value?

" *Those peculiarities which first appear, should be more
highly estimated than those which appear subsequently. A
system, in order to be true and natural, must agree with
the sequence of the organs in the development of the
embryo.*" If the earlier manifested peculiarities are to
be estimated more highly than those which afterwards
make their appearance, then in those cases in which
the structure of the adult animal requires one position
in the system, and that of the larva another, the
latter and not the former must decide the point. As
the *Lernœœ* and Cirripedes, on account of their Nauplius-
brood, were separated from their previous connexions
and referred to the Crustacea, we shall, for the same
reason, have to separate *Penëus* from the Prawns and
unite it with the Copepoda and Cirripedia. But the
most zealous embryomaniac would probably shrink
from this course.

A "true and natural system" of the Crustacea to
be in accordance with the sequence of the phenomena
would have to take into account in the first place the
various modes of segmentation, then the position of
the embryo, next the number of limbs produced within
the egg and so forth, and might be represented somewhat as follows :—

CLASSIS CRUSTACEA.

Sub-class I. Holoschista. — Segmentation complete. No primitive band. Nauplius-brood.

Ord. 1. *Ceratometopa.*—Nauplius with frontal horns. (Cirripedia, Rhizocephala.)

Ord. 2. Leiometopa.—Nauplius without frontal horns. (Copepoda, without *Achtheus*, &c., Phyllopoda, *Penëus.*)

Sub-class II. Hemischista.—Segmentation not complete.

A. Nototropa.—Embryo bent upwards.

Ord. 3. *Protura.*—The tail is first formed. (*Mysis.*)

Ord. 3. *Saccomorpha.*—A maggot-like larva-skin is first formed. (*Isopoda.*)

B. Gasterotropa.—Embryo bent ventrally.

Ord. 5. *Zoëogona.*—Full number of limbs not produced in the egg. Zoëa-brood. (The majority of the Podophthalmata.)

Ord. 6. *Ametabola.*—Full number of limbs produced in the egg. (*Astacus, Gecarcinus, Amphipoda* less *Hyperia?*)

This sample may suffice. The farther we go into details in this direction, the more brilliantly, as may easily be imagined, does the naturalness of such an arrangement as this force itself upon us.

All things considered, we may apply the judgment which Agassiz pronounced upon Darwin's theory, with far greater justice to the propositions just examined:— "No theory," says he, " however plausible it may be, can be admitted in science, unless it is supported by facts."

CHAPTER XI.

ON THE PROGRESS OF EVOLUTION.

From this scarcely unavoidable but unsatisfactory side-glance upon the old school, which looks down with so great an air of superiority upon Darwin's " intellectual dream " and the " giddy enthusiasm " of its friends, I turn to the more congenial task of considering the de-velopmental history of the Crustacea from the point of view of the Darwinian theory.

Darwin himself, in the thirteenth chapter of his book, has already discussed the conclusions derived from his hypotheses in the domain of developmental history. For a more detailed application of them, however, it is necessary in the first place to trace these general con-clusions a little further than he has there done.

The changes by which young animals depart from their parents, and the gradual accumulation of which causes the production of new species, genera, and families, may occur at an earlier or later period of life, —in the young state, or at the period of sexual ma-turity. For the latter is by no means always, as in the Insecta, a period of repose; most other animals even then continue to grow and to undergo changes.

(See above, the remarks on the males of the Amphipoda.) Some variations, indeed, from their very nature, can only occur when the young animal has attained the adult stage of development. Thus the Sea Caterpillars (*Polynoë*) at first possess only a few body-segments, which, during development, gradually increase to a number which is different in different species, but constant in the same species ; now before a young animal could exceed the number of segments of its parents, it must of course have attained that number. We may assume a similar supplementary progress wherever the deviation of the descendants consists in an addition of new segments and limbs.

Descendants therefore reach a new goal, either by deviating sooner or later whilst still on the way towards the form of their parents, or by passing along this course without deviation, but then, instead of standing still, advance still farther.

The former mode will have had a predominant action where the posterity of common ancestors constitutes a group of forms standing upon the same level in essential features, as the whole of the Amphipoda, Crabs, or Birds. On the other hand we are led to the assumption of the second mode of progress, when we seek to deduce from a common original form, animals some of which agree with young states of others.

In the former case the developmental history of the descendants can only agree with that of their ancestors up to a certain point at which their courses separate,— as to their structure in the adult state it will teach us nothing. *In the second case the entire development of the progenitors is also passed through by the descendants, and, therefore, so far as the production of a species depends upon this second mode of progress, the historical development of the species will be mirrored in its developmental history.* In the short period of a few weeks or months, the changing forms of the embryo and larvæ will pass before us, a more or less complete and more or less true picture of the transformations through which the species, in the course of untold thousands of years, has struggled up to its present state.

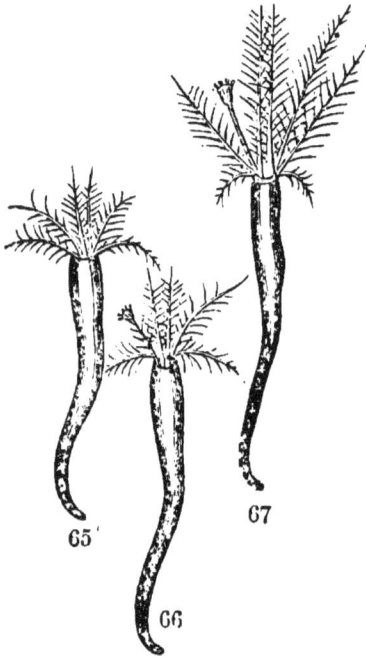

One of the simplest examples is furnished by the development of the Tubicolar Annelids; but from its very simplicity it appears well adapted to open the eyes of many who, perhaps, would rather

Figs. 65, 66, 67.[1]

[1] Figs. 65–67. Young Tubicolar worms, magnified with the simple lens about 6 diam.: 65.ª without operculum, *Protula*-stage ; 66. with a barbate opercular peduncle, *Filograna*-stage; 67. with a naked opercular peduncle, *Serpula*-stage.

ª Fig. 65 is drawn from memory, as the little animals, which I at first took for young *Protulæ*, only attracted my attention when I remarked the appearance of the operculum, which induced me to draw them.

not see, and it may therefore find a place here. Three years ago I found on the walls of one of my glasses some small worm-tubes (fig. 65), the inhabitants of which bore three pairs of barbate branchial filaments, and had no operculum. According to this we should have been obliged to refer them to the genus *Protula*. A few days afterwards one of the branchial filaments had become thickened at the extremity into a clavate operculum (fig. 66), when the animals reminded me, by the barbate opercular peduncle, of the genus *Filograna*, only that the latter possesses two opercula. In three days more, during which a new pair of branchial filaments had sprouted forth, the opercular peduncle had lost its lateral filaments (fig. 67), and the worms had become *Serpulæ*. Here the supposition at once presents itself that the primitive tubicolar worm was a *Protula*,—that some of its descendants, which had already become developed into perfect *Protulæ*, subsequently improved themselves by the formation of an operculum which might protect their tubes from inimical intruders,—and that subsequent descendants of these latter finally lost the lateral filaments of the opercular peduncle, which they, like their ancestors, had developed.

What say the schools to this case? Whence and for what purpose, if the *Serpulæ* were produced or created as ready-formed species, these lateral filaments of the opercular peduncle? To allow them to sprout forth merely for the sake of an invariable plan of structure, even when they must be immediately re-

I

tracted again as superfluous, would certainly be an
evidence rather of childish trifling or dictatorial pe-
dantry, than of infinite wisdom. But no, I am mistaken; from the beginning of all things the Creator
knew, that one day the inquisitive children of men
would grope about after analogies and homologies, and
that Christian naturalists would busy themselves with
thinking out his Creative ideas; at any rate, in order
to facilitate the discernment by the former that the
opercular peduncle of the *Serpulæ* is homologous with
a branchial filament, He allowed it to make a *détour* in
its development, and pass through the form of a barbate branchial filament.

The historical record preserved in developmental his-
tory is gradually EFFACED *as the development strikes into*
a constantly straighter course from the egg to the perfect
animal, and it is frequently SOPHISTICATED *by the struggle*
for existence which the free-living larvæ have to undergo.

Thus as the law of inheritance is by no means strict,
as it gives room for individual variations with regard
to the form of the parents, this is also the case with
the succession in time of the developmental processes.
Every father of a family who has taken notice of such
matters, is well aware that even in children of the same
parents, the teeth, for example, are not cut or changed,
either at the same age, or in the same order. Now in
general it will be useful to an animal to obtain as
early as possible those advantages by which it sustains
itself in the struggle for existence. A precocious appearance of peculiarities originally acquired at a later

period will generally be advantageous, and their re-
tarded appearance disadvantageous; the former, when
it appears accidentally, will be preserved by natural
selection. It is the same with every change which
gives to the larval stages, rendered multifarious by
crossed and oblique characters, a more straightforward
direction, simplifies and abridges the process of development, and forces it back to an earlier period of life,
and finally into the life of the egg.

As this conversion of a development passing through
different young states into a more direct one, is not the
consequence of a mysterious inherent impulse, but de-
pendent upon advances accidentally presenting themselves, it may take place in the most nearly allied
animals in the most various ways, and require very
different periods of time for its completion. There is
one thing, however, that must not be overlooked here.
The historical development of a species can hardly
ever have taken place in a continuously uniform flow;
periods of rest will have alternated with periods of
rapid progress. But forms, which in periods of rapid
progress were severed from others after a short duration, must have impressed themselves less deeply upon
the developmental history of their descendants, than
those which repeated themselves unchanged, through a
long series of successive generations in periods of rest.
These more fixed forms, less inclined to variation, will
present a more tenacious resistance in the transition
to direct development, and will maintain themselves
in a more uniform manner and to the last, however

different may be the course of this process in other respects.

In general, as already stated, it will be advantageous to the young to commence the struggle for existence in the form of their parents and furnished with all their advantages—in general, but not without exceptions. It is perfectly clear that a brood capable of locomotion is almost indispensable to attached animals, and that the larvæ of sluggish Mollusca, or of worms burrowing in the ground, &c., by swarming briskly through the sea perform essential services by dispersing the species over wider spaces. In other cases a metamorphosis is rendered indispensable by the circumstance that a division of labour has been set up between the various periods of life ; for example, that the larvæ have exclusively taken upon themselves the business of nourishment. A further circumstance to be taken into consideration is the size of the eggs,—a simpler structure may be produced with less material than a more compound one,—the more imperfect the larva, the smaller may the egg be, and the larger is the number of these that the mother can furnish with the same expenditure of material. As a rule, I believe indeed, this advantage of a more numerous brood will not by any means outweigh that of a more perfect brood, but it will do so in those cases in which the chief difficulty of the young animals consists in finding a suitable place for their development, and in which, therefore, it is of importance to disperse the greatest possible number of germs, as in many parasites.

As the conversion of the original development with metamorphosis into direct development is here under discussion, this may be the proper place to say a word as to the already indicated absence of metamorphosis in fresh-water and terrestrial animals the marine allies of which still undergo a transformation. This circumstance seems to be explicable in two ways. Either species without a metamorphosis migrated especially into the fresh waters, or the metamorphosis was more rapidly got rid of in the emigrants than in their fellows remaining in the sea.

Animals without a metamorphosis would naturally transfer themselves more easily to a new residence, as they had only themselves and not at the same time multifarious young forms to adapt to the new conditions. But in the case of animals with a metamorphosis, the mortality among the larvæ, always considerable, must have become still greater under new than under accustomed conditions, every step towards the simplification of the process of development must therefore have given them a still greater preponderance over their fellows, and the effacing of the metamorphosis must have gone on more rapidly. What has taken place in each individual case, whether the species has immigrated after it had lost the metamorphosis, or lost the metamorphosis after its immigration, will not always be easy to decide. When there are marine allies without, or with only a slight metamorphosis, like the Lobster as the cousin of the Cray-fish, we may take up the former supposition; when allies with a metamorphosis still live upon the land or in fresh water, as in the case of *Gecarcinus*, we may adopt the latter.

That besides this gradual extinction of the primitive history, a *falsification* of the record preserved in the developmental history takes place by means of the struggle for existence which the free-living young states have to undergo, requires no further exposition. For it is perfectly evident that the struggle for existence and natural selection combined with this, must act in the same way, in change and development, upon larvæ which have to provide for themselves, as upon adult animals. The changes of the larvæ, independent of the progress of the adult animal, will become the more considerable, the longer the duration of the life of the larva in comparison to that of the adult animal, the greater the difference in their mode of life, and the more sharply marked the division of labour between the different stages of development. These processes have to a certain extent an action opposed to the gradual extinction of the primitive history; they increase the differences between the individual stages of development, and it will be easily seen how even a straightforward course of development may be again converted by them into a development with metamorphosis. By this means many, and it seems to me valid reasons may be brought up in favour of the opinion that the most ancient Insects approached more nearly to the existing Orthoptera, and perhaps to the wingless Blattidæ, than to any other order, and that the " complete metamorphosis" of the Beetles, Lepidoptera, &c., is of later origin. There were, I believe, perfect Insects before larvæ and pupæ; but, on the contrary, Nauplii and Zoëæ far earlier than perfect Prawns. In contradistinction to the *inherited* metamorphosis of the Prawns, we may call that of the Coleoptera, Lepidoptera, &c., an *acquired* metamorphosis.[2]

[2] I will here briefly give my reasons for the opinion that the so-called "complete metamorphosis" of Insects, in which these animals quit the egg as grubs or caterpillars, and afterwards become quiescent pupæ incapable of feeding, was not inherited from the primitive ancestor of all Insects, but acquired at a later period.

The order Orthoptera, including the Pseudoneuroptera (*Ephemera, Libellula,* &c.) appears to approach nearest to the primitive form of Insects. In favour of this view we have :—

1. The structure of their buccal organs, especially the formation of the labium, "which retains, either perfectly or approximately, the original form of a second pair of maxillæ" (Gerstäcker).

2. The segmentation of the abdomen; "like the labium, the abdomen also very generally retains its original segmentation, which is shown in the development of eleven segments" (Gerstäcker). The Orthoptera with eleven segments in the abdomen, agree perfectly in the number of their body-segments with the Prawn-larva represented in fig. 33, or indeed, with the higher Crustacea (Podophthalma and Edriophthalma) in general, in which the historically youngest last thoracic segment (see p. 123), which is sometimes late-developed, or destitute of appendages, or even deficient, is still wanting.

3. That, as in the Crustacea, the sexual orifice and anus are placed upon different segments; "whilst the former is situated in the ninth segment, the latter occurs in the eleventh" (Gerstäcker).

4. Their palæontological occurrence; "in a fossil state the Orthoptera make their appearance the earliest of all Insects, namely as early as the Carboniferous formation, in which they exceed all others in number" (Gerstäcker).

5. The absence of uniformity of habit at the present day in an order so small when compared with the Coleoptera, Hymenoptera, &c. For this also is usually a phenomenon characteristic of very ancient groups of forms which have already overstepped the climax of their development, and is explicable by extinction in mass. A Beetle or a Butterfly is to be recognised as such at the first glance, but only a thorough

Which of the different modes of development at present occurring in a class of animals may claim to be

investigation can demonstrate the mutual relationships of *Termes*, *Blatta*, *Mantis*, *Forficula*, *Ephemera*, *Libellula*, &c. I may refer to a corresponding remarkable example from the vegetable world: amongst Ferns the genera *Aneimia*, *Schizæa* and *Lygodium*, belonging to the group *Schizæaceæ* which is very poor in species, differ much more from each other than any two forms of the group *Polypodiaceæ* which numbers its thousands of species.

If, from all this, it seems right to regard the Orthoptera as the order of Insects approaching most nearly to the common primitive form, we must also expect that their mode of development will agree better with that of the primitive form, than, for example, that of the Lepidoptera, in the same way that some of the Prawns (*Penëus*) approaching most closely the primitive form of the Decapoda, have most truly preserved their original mode of development. Now, the majority of the Orthoptera quit the egg in a form which is distinguished from that of the adult Insect almost solely by the want of wings; these larvæ then soon acquire rudiments of wings, which appear more strongly developed after every moult. Even this perfectly gradual transition from the youngest larva to the sexually mature Insect, preserves in a far higher degree the picture of an original mode of development, than does the so-called complete metamorphosis of the Coleoptera, Lepidoptera, or Diptera, with its abruptly separated larva-, pupa- and imago-states.

The most ancient Insects would probably have most resembled these wingless larvæ of the existing Orthoptera. The circumstance that there are still numerous wingless species among the Orthoptera, and that some of these (*Blattidæ*) are so like certain Crustacea (Isopods) in habit that both are indicated by the same name ("*Baratta*") by the people in this country, can scarcely be regarded as of any importance.

The contrary supposition that the oldest Insects possessed a "complete metamorphosis," and that the "incomplete metamorphosis" of the Orthoptera and Hemiptera is only of later origin, is met by serious difficulties. If all the classes of Arthropoda (Crustacea, Insecta, Myriopoda and Arachnida) are indeed all branches of a common stem (and of this there can scarcely be a doubt), it is evident that the water-inhabiting and water-breathing Crustacea must be regarded as the original stem from which the other terrestrial classes, with their tracheal respiration, have branched off. But nowhere among the Crustacea is there a mode of development comparable to the "complete metamorphosis" of the Insecta, nowhere among the young or adult that approaching most nearly to the original one, is easy to judge from the above statements.

The primitive history of a species will be preserved in its developmental history the more perfectly, the longer the series of young states through which it passes by uniform steps ; and the more truly, the less the mode of life of the young departs from that of the adults, and the less the peculiarities of the individual young states can be conceived as transferred back from later ones in previous periods of life, or as independently acquired.

Let us apply this to the Crustacea.

Crustacea are there forms which might resemble the maggots of the Diptera or Hymenoptera, the larvæ of the Coleoptera, or the caterpillars of the Lepidoptera, still less any bearing even a distant resemblance to the quiescent pupæ of these animals. The pupæ, indeed, cannot at all be regarded as members of an original developmental series, the individual stages of which represent permanent ancestral states, for an animal like the mouthless and footless pupa of the Silkworm, enclosed by a thick cocoon, can never have formed the final, sexually mature state of an Arthropod.

In the development of the Insecta we never see new segments added to those already present in the youngest larvæ, but we do see segments which were distinct in the larva afterwards become fused together or disappear. Considering the parallelism which prevails throughout organic nature between palæontological and embryonic development, it is therefore improbable that the oldest Insects should have possessed fewer segments than some of their descendants. But the larvæ of the Coleoptera, Lepidoptera, &c., never have more than nine abdominal segments, it is therefore not probable that they represent the original young form of the oldest Insects, and that the Orthoptera, with an abdomen of eleven segments, should have been subsequently developed from them.

Taking into consideration on the one hand these difficulties, and on the other the arguments which indicate the Orthoptera as the order most nearly approaching the primitive form, it is my opinion that the " incomplete metamorphosis " of the Orthoptera is the primitive one, *inherited* from the original parents of all Insects, and the " complete metamorphosis" of the Coleoptera, Diptera, &c., a subsequently *acquired* one.

CHAPTER XII.

PROGRESS OF EVOLUTION IN CRUSTACEA.

According to all the characters established in the last paragraph, the Prawn that we traced from the Nauplius through states analogous to Zoëa and *Mysis* to the form of a Macrurous Crustacean appears at present to be the animal, which in the section of the higher Crustacea (Malacostraca) furnishes the truest and most complete indications of its primitive history. That it is the most complete is at once evident. That it is the truest must be assumed, in the first place, because the mode of life of the various ages is less different than in the majority of the other Podophthalma; for from the Nauplius to the young Prawn they were found swimming freely in the sea, whilst Crabs, *Porcellanæ*, the Tatuira, *Squilla*, and many Macrura, when adult usually reside under stones, in the clefts of rocks, holes in the earth, subterranean galleries, sand, &c., not to mention other deviations in habits such as are presented by the Hermit Crabs, *Pinnotheres*, &c.,—and secondly and especially because the peculiarities which distinguish the Zoëa of this species particularly from other Zoëæ (the employment of the anterior limbs for swimming, the furcate tail, the simple heart, the deficiency of the paired eyes

and abdomen at first, &c.) are neither to be deduced
from a retro-transfer of late-acquired advantages to this
early period of life, nor to be regarded at all as advantages over other Zoëæ which the larva might have
acquired in the struggle for existence.

A similar development must have been once passed
through by the primitive ancestor of all Malacostraca,
probably differing from that of our Prawn, especially in
the circumstance that it would go on more uniformly
without the sudden change of form and mode of locomotion produced in the latter by the simultaneous sprouting forth and entering into action in the Nauplius of
four and in the Zoëa of five pairs of limbs. It is to be
supposed that, not only originally but even still, in the
larvæ of the first Malacostraca, the new body-segments
and pairs of limbs are formed singly,—first of all the
segments of the fore-body, then those of the abdomen,
and finally those of the middle-body,—and, moreover,
that in each region of the body the anterior segments
were formed earlier than the posterior ones, and therefore last of all the hindermost segment of the middle-
body. Of this original mode more or less distinct traces
still remain, even in species in which, in other respects,
the course of development of their ancestors is already
nearly effaced. Thus the abdominal feet of the Prawn-
larva represented in fig. 33, are formed singly from
before backwards, and after these the last feet of the
middle-body; thus, in *Palinurus*, the last two pairs of
feet of the middle-body are formed later than the rest;
thus in the young larvæ of the Stomapoda the last

three abdominal segments are destitute of limbs, which
are still wanting on the last of them in older larvæ ;
and thus, in the Isopoda, the historically newest pair of
feet is produced later than all the rest. In the Copepoda this formation of new segments and limbs, gradually advancing from before backwards, is more perfectly
preserved than in any of the higher Crustacea.[1]

The original development of the Malacostraca starting from the Nauplius, or the lowest free-living grade
with which we are acquainted in the class of Crustacea,
is now-a-days nearly effaced in the majority of them.
That this extinction has actually taken place in the way
already deduced as a direct consequence from Darwin's
theory, will be the more easily demonstrated, the more
this process is still included in the course of life, and
the less completely it is already worn out. We may
hope to obtain the most striking examples in the still
unknown developmental history of the various Schizo-
poda, Peneïdæ, and, indeed, of the Macrura in general.
At present the multifarious Zoëa-forms appear to be

---

[1] It is well known that, in many cases, even in adult animals the last
segment of the middle-body, or some of its last segments, either want
their limbs or are themselves deficient (*Entoniscus Porcellanæ ♂, Leucifer,*
&c.). This might be due to the animals having separated from the
common stem before these limbs were formed at all. But in those
cases with which I am best acquainted, it seems to me more probable
that the limbs have been subsequently lost again. That these particular
limbs and segments are more easily lost than others is explained by the
circumstance that, as the youngest, they have been less firmly fixed by
long-continued inheritance. ("Mr. Dana believes, that in ordinary
Crustaceans, the abortion of the segments with their appendages almost
always takes place at the posterior end of the cephalothorax."—Darwin,
Balanidæ, p. 111.)

particularly instructive. Almost all the peculiarities by which they depart from the primitive form of the Zoëa of *Penĕus* (figs. 29, 30, 32), may in fact be conceived as transferred back from a later period into this early period of life. This is the case with the large compound eyes,—with the structure of the heart,—with the raptorial feet in *Squilla*,—and with the powerful, muscular, straightly-extended abdomen in *Palæmon*, *Alpheus*, *Hippolyte*, and the Hermit Crabs. (In the latter, indeed, the abdomen of the adult animal is a shapeless sac filled with the liver and generative organs, but it is still tolerably powerful in the *Glaucothoë*-stage, and was certainly still more powerful when this stage was still the permanent form of the animal.) It is also the case with the abdomen of the Zoëæ of the Crabs, the *Porcellanæ*, and the Tatuira, which is still powerful, although usually bent under the breast; the two last swim tolerably by means of the abdomen, even when adult, as do the true Crabs in the young state known as *Megalops*. It is the case, lastly, with the conversion of the two anterior pairs of limbs into antennæ. The second pair of antennæ, which, in the various Zoëæ always remains a step behind that of the adult animal, is particularly remarkable. In the Crabs the "scale" is entirely wanting; their Zoëæ have it indicated in the form of a moveable appendage, which is often exceedingly minute. In the Hermit Crabs a similar, usually moveable, spiniform process occurs as the remains of the scale; their Zoëæ have a well-developed but inarticulate scale. A precisely similar scale is possessed by the adult Prawns, in the Zoëæ of which it exists still in a jointed form, like the outer branch of the second pair of feet of the Nauplius or *Penëus*-Zoëa.

The long, spiniform processes on the carapace of the Zoëæ of the Crabs and *Porcellanæ* are not to be explained in this way, but their advantage to the larvæ is evident. Thus, for example, if the body of the Zoëa of *Porcellana stellicola* (fig. 24), without the processes of the carapace and without the abdomen, which however is not rigidly extensible, is scarcely half a line in length, whilst with the processes it is four lines long, a mouth of eight times the width is necessary in order to swallow the little animal when thus armed.[2] Consequently these processes of the carapace may be regarded as acquired by the Zoëa itself in the struggle for existence.

The formation of new limbs beneath the skin of the larvæ is also to be referred to an earlier occurrence of processes which originally took place at a later period. The original course must have been that they sprouted forth in a free form upon the ventral surface of the larva in the next stage after the change of skin; whilst now they are developed before the change of skin, and thus only come into action a stage earlier. In larvæ which, for other reasons, must be regarded as more nearly approaching the primitive form, the original

---

[2] *Persephone*, a rare Crab, belonging to the family Leucosiidæ, is served in the same manner by its long chelate feet. If we seize the animal, it extends them most obstinately straight downwards, so that in all probability we should more easily break than bend them.

mode usually prevails in this particular also. Thus the caudal feet (the "lateral caudal lamellæ") are formed freely on the ventral surface in *Euphausia* and the Prawns with Nauplius-brood, and within the caudal lamellæ in the Prawns with Zoëa-brood, in *Pagurus* and *Porcellana*.

A compression of several stages into one, and thereby an abridgement and simplification of the course of development, is expressed in the simultaneous appearance of several new pairs of limbs.

How earlier young states may gradually be completely lost, is shown by *Mysis* and the Isopoda. In *Mysis* there is still a trace of the Nauplius-stage; being transferred back to a period when it had not to provide for itself, the Nauplius has become degraded into a mere skin; in *Ligia* (figs. 36, 37) this larva-skin has lost the last traces of limbs, and in *Philoscia* (fig. 38) it is scarcely demonstrable.

Like the spinous processes of the Zoëæ, the chelæ on the penultimate pair of feet of the young *Brachyscelus* are to be regarded as acquired by the larva itself.˙ The adult animals swim admirably and are not confined to their host; as soon as the specimens of *Chrysaora Blossevillei*, Less., or *Rhizostoma cruciatum*, Less., on which they are seated, become the sport of the waves in the neighbourhood of the shore, they escape from them, and are only to be obtained from lively Acalephs. The young are helpless creatures and bad swimmers; a special apparatus for adhesion must be of great service to them.

To review the developmental history of the different
Malacostraca in detail would furnish no results at
all correspondent to the time occupied by it,—if our
knowledge was more complete it would be more profit-
able. I therefore abandon it, but will not omit to
mention that in it many difficulties which cannot at
present be satisfactorily solved would present themselves. To these isolated difficulties I ascribe the less
importance, however, because even a little while ago,
before the discovery of the Prawn-Nauplius, this entire
domain of the development of the Malacostraca was
almost inaccessible to Darwin's theory.

Nor will I dwell upon the contradictions which appear
to result from the application of the Darwinian theory
to this department. I leave it to our opponents to
find them out. Most of them may easily be proved to
be only apparent. There are two of these objections,
however, which lie so much on the surface that they
can hardly escape being brought forward, and these, I
think, I must get rid of.

"The peculiarities in which the Zoëæ of the Crabs,
the *Porcellanæ*, the Tatuira, the Hermit Crabs, and the
Prawns with Zoëa-brood agree, and by which they are
in common distinguished from the larvæ of *Penëus*
produced from Nauplii, forces us (it might be said) to
the supposition that the common ancestor of these
various Decapods quitted the egg in a similar Zoëa-
form. But then neither *Penëus* with its Nauplius-
brood, nor even apparently the *Palinuri* could be referred back to this ancestor. The mode of development

of *Penëus* and *Palinurus*, as also several peculiar larvæ of unknown origin, but which are in all probability to be attributed to Macrurous Crustacea, necessitate on the contrary the opposite supposition, namely, that the different groups of the Macrura have passed from their original to their present mode of development independently of each other and also independently of the Crabs." To this we may answer that the occurrence of the Zoëa-form in all the above-mentioned Decapoda, its existence in *Penëus* during the whole of that period of life which is richest in progress and in which the wide gap between the Nauplius and the Decapod is filled up, its recurrence even in the development of the Stomapoda, the occurrence of a larval form closely approaching the youngest Zoëa of *Penëus* in the Schizopod genus *Euphausia*, and the reminiscence of the structure of Zoëa, which even the adult *Tanais* has preserved in its mode of respiration,—all indicate Zoëa as one of those steps in development which persisted as a permanent form throughout a long period of repose, perhaps through a whole series of geological formations, and thus has also made a deeper impression upon the development of its descendants, and formed a firmer nucleus in the midst of other and more readily effaced young states. It cannot, therefore, surprise us that in transitions from the original mode of metamorphosis to direct development, even when produced independently, the larval life commences in the same way with this Zoëa-form in different families, in which the earlier stages of development are effaced. But except what is common to

K

all Zoëæ, and what may easily be explained as being transferred back from a later into this stage, the Zoëæ of the
Crabs, for example, agree with those of *Pagurus* and
Palæmon in no single peculiarity of structure which
leads us to suppose a common inheritance. Consequently we may apparently assume, without hesitation,
that when the Brachyura and Macrura separated, the
primitive ancestors of each of these groups passed
through a more complete metamorphosis, and that the
transition to the present mode of development belongs
to a later period. With regard to the Brachyura, it
may be added that in them this transition occurred
only a little later and indeed before the existing families
separated. The arrangement of the processes of the
carapace, and, still more, the similar number of the
caudal setæ in the most different Zoëæ of Crabs (figs.
19–23) prove this. Such an accordance in the number
of organs apparently so unimportant is only explicable
by common inheritance. We may predict with certainty that amongst the Brachyura no species will
occur which, like *Penëus*, still produces Nauplius-
brood.[3]

As we have already seen, *Mysis* and the Isopoda
depart from all other Crustacea very remarkably by
the fact that their embryos are curved upwards, instead

---

[3] I must not omit remarking that what has been said as to the
development of the Crabs applies essentially only to the groups
Cyclometopa, Catometopa and *Oxyrhyncha*, placed together by Alph.
Milne-Edwards as " Eustomés." Among the *Oxystomata*, as also among
the " Anomura apterura," Edw., which approach so nearly to the Crabs,
I am unacquainted with the earliest young states of any of the species.

of, as elsewhere, downwards.   Does not so isolated a
phenomenon as this, it might be asked, in the sense of
Darwin's theory, indicate a common inheritance ?   Does
it not necessitate that we should unite as the descendants of the same primitive ancestors, *Mysis* with the
Isopoda on the one hand, and on the other the rest of
the Podophthalma with the Amphipoda?   I think not.
Such a necessity exists only for those who estimate a
peculiarity at a higher value because it makes its appearance at an earlier period of the egg-life.   Whoever
regards species as not created independently and un-
changeably, but as having gradually become what they
are, will say to himself that, when the ancestors of our
Mysides came (probably much later than those of the
Amphipoda and Isopoda) to develope numerous body-
segments and limbs whilst still embryos, as they could
no longer find room in the egg when extended straight
out, and were therefore compelled to bend themselves,
this could only take place either upwards or downwards, and whatever conditions may have decided the
direction actually adopted, any near relationship to
either of the two orders of Edriophthalma could hardly
have taken part in it.

It may, however, be remarked, that the different curvature of the embryo in the Amphipoda and Isopoda is so
far instructive, as it proves that their present mode of development was adopted only after the separation of these
orders, and that, in the primitive stock of the Edriophthalma, the embryos were, if not Nauplii, at least short
enough in the body to find room in the egg in an

extended position, like the larvæ of *Achtheres* enclosed by the Nauplius-skin. On the other hand the uniformity of development that prevails in each of the two orders—which is expressed in the Amphipoda for example in the formation of the "micropylar apparatus," in the Isopoda in the want of the last pair of ambulatory feet—testifies that the present mode of development has come down from a very early period and extends back beyond the separation of the present families. In these two orders also, as well as in the Crabs, we can hardly hope to find traces of earlier young states, unless it be in the family of the Tanaidæ.[4] If any one will furnish me with an Amphipod or an Isopod with Nauplius-brood, the existence of which would not be more remarkable in independently produced species than that of a Prawn with Nauplius-brood, I will abandon the whole Darwinian theory.

With regard to the Crabs, and also to the Isopoda and Amphipoda, we were led to the assumption, that, about the period when these groups started from the

[4] Whether the want of the abdominal feet in the young of *Tanais* be an inheritance from the time of the primitive Isopoda, or a subsequently acquired peculiarity, which appears to me the more admissible view at present, may perhaps be decided with some certainty, when we become acquainted with the development and mode of life of its family allies, *Apseudes* and *Rhœa*. The latter, as is well known, is the only Isopod which possesses a secondary flagellum on the anterior antennæ. I have recently obtained a new and unexpected proof that the *Tanaidæ* ("Asellotes hétéropodes" M.-Edw.) of all known Crustacea approach most closely to the primitive form of the Edriophthalma. Mr. C. Spence Bate writes to me: "*Apseudes*, as far as I know, is the *only* Isopod in which the antennal scale so common in the Macrura is present on the lower antennæ."

common stem, a simplification of their process of development took place. This also seems to be intelligible from Darwin's theory. When any circumstances favourable to a group of animals caused its wider diffusion and divergence into forms adapting themselves to new and various conditions of existence, this greater variability, which betrays itself in the production of new forms, will also favour the simplification of the development which is almost always advantageous, and moreover, exactly at this period, during adaptation to new circumstances, as has already been indicated with regard to fresh-water animals, this simplification will be doubly beneficial, and therefore, in connexion with this, a doubly strict selection will take place.

So much for the development of the higher Crustacea.

A closer examination of the developmental history of the lower Crustacea is unnecessary after what has been said in general upon the historical significance of the young states, and the application of this which has just been made to the Malacostraca. We may see, without further discussion, how the representation given by Claus of the development of the Copepoda may pass almost word for word as the primitive history of those animals; we may find in the Nauplius-skin of the larvæ of *Achtheres* and in the egg-like larva of *Cryptophialus*, precisely similar traces of a transition towards direct development, as were presented by the *Nauplius*-envelope of the embryos of *Mysis* and the maggot-like larva of *Ligia*, &c.

It will be sufficient to indicate an essential difference

in the process of development in the higher and lower
Crustacea. In the latter all new body-segments and
limbs which insert themselves between the two terminal regions of the Nauplius, are formed in uninterrupted
sequence from before backwards; in the former there is
further a new formation in the middle of the body (the
middle-body), which pushes itself in between the fore-
body and the abdomen in the same way, as these have
done on their part between the head and tail of the
Nauplius. Thus, that which appears probable even
from the comparison of the limbs of the adult animal,
finds fresh support in the developmental history, namely,
that the lower Crustacea, like the Insects, are entirely
destitute of the region of the body corresponding to
the middle-body of the Malacostraca. It seems probable that the swimming feet of the Copepoda, as also
of the pupæ of Cirripedia and Rhizocephala, represent
the abdominal feet of the Malacostraca, that is to say,
are derived by inheritance from the same source with
them.

It would be easy to weave together the separate
threads furnished by the young forms of the various
Crustacea, into a general picture of the primitive history of this class. Such a picture, drawn with a little
skill, and finished in lively colours, would certainly
be more attractive than the dry discussions which I
have tacked on to the developmental history of these
animals. But the mode of weaving in the loose threads
would still in many cases be arbitrary, and to be
effected with equal justice in various ways; and many

gaps would still have to be filled up by means of more or less bold assumptions. Those who have not wandered much in this region of research would then readily believe that they were standing upon firm ground, where mere fancy had thrown an airy bridge ; those acquainted with the subject, on the other hand, would soon find out these weak points in the structure, but would then be easily led to regard even what was founded upon well considered facts, as merely floating in the air. To obviate these misconceptions of its true contents from either side, it would be necessary to accompany such a picture throughout with lengthy, dry explanations. This has deterred me from further filling in the outline which I had already sketched.

I will only give, as an example, the probable history of the production of a single group of Crustacea, and indeed of the most abnormal of all, the RHIZOCEPHALA, which in the sexually mature state differ so enormously even from their nearest allies, the Cirripedia, and from their peculiar mode of nourishment stand quite alone in the entire animal kingdom.

I must preface this with a few words upon the homology of the roots of the Rhizocephala, i.e. the tubules which penetrate from its point of adhesion into the body of the host, ramify amongst the viscera of the latter, and terminate in cæcal branchlets. In the pupæ of the Rhizocephala (fig. 58) the foremost limbs ("prehensile antennæ") bear, on each of the two terminal joints, a tongue-like, thin-skinned appendage, in which we may generally observe a few small strongly refractive gra-

nules, like those seen in the roots of the adult animal.
I have therefore supposed these appendages to be the
rudiments of the future roots. A perfectly similar
appendage, " a most delicate tube or ribbon," was
found by Darwin in free-swimming pupæ of *Lepas
australis* on the last joints of the " prehensile antennæ."
From the perfect accordance in their entire structure
shown by the pupæ of the Rhizocephala and Cirripedia,
there can be no doubt that the appendages of *Sacculina*
and *Lepas*, which are so like each other and spring from
the same spot, are homologous structures.

Now in three species of *Lepas*, in *Dichelaspis War-
wickii* and in *Scalpellum Peronii*, Darwin saw, on tear-
ing recently-affixed animals from their point or support,
that a long narrow band issued from the same point of
the antennæ; its end was torn away, and in *Dichelaspis*,
judging from its ragged appearance, it had attached
itself firmly to the support. From this it follows that
this appendage in *Lepas australis* can hardly be anything but a young cement-duct. If, therefore, the
supposition that the appendages on the antennæ of
the pupæ of Rhizocephala are young roots be cor-
rect, the roots of the Rhizocephala are homologous
with the cement-ducts of the Cirripedia. And this,
strange as it may appear at the first glance, seems to me
scarcely doubtful. It is true that the act of adhesion
of the Rhizocephala has never yet been observed, but it
is more than probable that they attach themselves, just
like the Cirripedia, by means of the antennæ, and that
therefore the points of attachment in the two groups

indicate homologous parts of the body.   From the point of attachment in the Rhizocephala the roots penetrate into the body of the host, whilst in the Cirripedia, the cement-ducts issue from the same point.   The roots are blind tubes, ramified in different ways in different species.   The cement-ducts in the basis of the Balanidæ likewise constitute a generally remarkably complicated system of ramified tubes, with regard to the mode of termination of which nothing certain has yet been made out.   Individual cæcal branches are not unfrequently seen even in the vicinity of the carina; and, at least in some species, in which the cement-ducts divide into extremely numerous and fine branchlets, forming a network which gradually becomes denser towards the circumference of the basis, these seem nowhere to possess an orifice.

Now as to the question: How were Cirripedia converted by natural selection into Rhizocephala?

A considerable number of existing Cirripedia settle exclusively or chiefly upon living animals;—on Sponges, Corals, Mollusks, Cetaceans, Turtles, Sea-Snakes, Sharks, Crustaceans, Sea Urchins, and even on Acalephs.   *Dichelaspis Darwinii* was found by Filippi in the branchial cavity of *Palinurus vulgaris*, and I have met with another species of the same genus in the branchial cavity of *Lupea diacantha*.

The same thing may have taken place in primitive times.   The supposition that certain Cirripedes might once upon a time have selected the soft ventral surface of a Crab, *Porcellana* or *Pagurus*, for its dwelling-place, has certainly nothing improbable about it. If then the cement-ducts of such a Cirripede instead of merely spreading on the surface, pierced or pushed before them the soft ventral skin and penetrated into the interior of the host, this must have been beneficial to the animal, because it would be thereby more securely attached and protected from being thrown off during the moulting of its host. Variations in this direction were preserved as advantageous.

But as soon as the cement-ducts penetrated into the body-cavity of the host and were bathed by its fluids, an endosmotic interchange must necessarily have been set up between the materials dissolved in these fluids and in the contents of the cement-ducts, and this interchange could not be without influence upon the nourishment of the parasite. The new source of nourishment opened up in this manner was, as constantly flowing, more certain than that offered by the nourishment accidentally whirled into the mouth of the sedentary animal. The individuals favoured in the development of the cement-ducts now converted into nutriferous roots, had more than others the prospect of abundant food, of vigorous growth, and of producing a numerous progeny. With the further development, assisted by natural selection, of the roots embracing the intestine of the host and spreading amongst its hepatic tubes, the introduction of nourishment through the mouth and all the parts implicated in it, such as the whirling cirri, the buccal organs, and the intestine, gradually lost their importance, became aborted by disuse, and finally disappeared without leaving a trace of their existence. Protected by the abdomen of the Crab, or by the shell inhabited by the *Pagurus*, the parasite also no longer required the calcareous test, in which, no doubt, the first Cirripedes settling upon these Decapods rejoiced. This protective covering, having become superfluous, also disappeared, and there remained at last only a soft sack filled with eggs, without limbs, without mouth or alimentary canal, and nourished, like a plant, by means of roots, which it pushed into the body of its host. The Cirripede had become a Rhizocephalon.

If it be desired to form a notion of what our parasite may have looked like when half way in its progress from the one form to the other, we may consult the figures given by Darwin, (Lepadidæ Pl., iv., figs. 1-7) of *Anelasma squalicola*. This Lepadide, which lives upon Sharks in the North Sea, seems, in fact, to be in the best way to lose its cirri and buccal organs in the same manner. The widely-cleft, shell-less test is supported upon a thick peduncle, which is immersed in the skin of the Shark. The surface of the peduncle is beset with much-ramified, hollow filaments, which " penetrate the Shark's flesh like roots " (Darwin). Darwin looked in vain for cement-glands and cement. It seems to me hardly doubtful, that the ramified hollow filaments are themselves nothing but the cement-ducts converted into nutritive roots, and that it is just in consequence of the development of this new source of nourishment, that the cirri and buccal organs are in the highest degree aborted. All the parts of the mouth are extremely

minute; the palpi and exterior maxillæ have almost
disappeared; the cirri are thick, inarticulate, and destitute of bristles; and the muscles both of the mouth and
cirri are without transverse striation. Darwin found
the stomach perfectly empty in the animal examined
by him.

Having reached the Nauplius, the extreme outpost of
the class, retiring furthest into the gray mist of primitive time, we naturally look round us to see whether
ways may not be descried thence towards other border-
ing regions. By the structure of the abdomen in
Nauplius we might be reminded, like Oscar Schmidt,
of the moveable caudal fork of the Rotatoria, which
many regard as near allies of the Crustacea, or at any
rate of the Arthropoda; in the six feet surrounding
the mouth we might imagine an originally radiate
structure, and so forth. But I can see nothing certain.
Even towards the nearer provinces of the Myriopoda
and Arachnida I can find no bridge. For the Insecta
alone, the development of the Malacostraca may perhaps present a point of union. Like many Zoëæ, the
Insecta possess three pairs of limbs serving for the
reception of nourishment, and three pairs serving for
locomotion; like the Zoëæ they have an abdomen without appendages; as in all Zoëæ the mandibles in
Insects are destitute of palpi. Certainly but little in
common, compared with the much which distinguishes
these two animal-forms. Nevertheless the supposition
that the Insecta had for their common ancestor a Zoëa

which raised itself into a life on land, may be recommended for further examination.

Much in what has been adduced above may be erroneous, many an interpretation may have failed, and many a fact may not have been placed in its proper light. But in one thing, I hope, I have succeeded,—in convincing *unprejudiced* readers, that Darwin's theory furnishes the key of intelligibility for the developmental history of the Crustacea, as for so many other facts inexplicable without it. The deficiencies of this attempt, therefore, must not be laid to the charge of the plan drawn out by the sure hand of the master, but solely to the clumsiness of the workman, who did not know how to find the proper place for every portion of his material.

INDEX.

LONDON: PRINTED BY W. CLOWES AND SONS, DUKE STREET, STAMFORD STREET, AND CHARING CROSS.

www.ingramcontent.com/pod-product-compliance
Lightning Source LLC
Chambersburg PA
CBHW021813190326
41518CB00007B/576